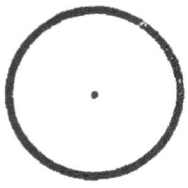

Chromatic Equivalents.
Fig. 2. Exp. XXIII.

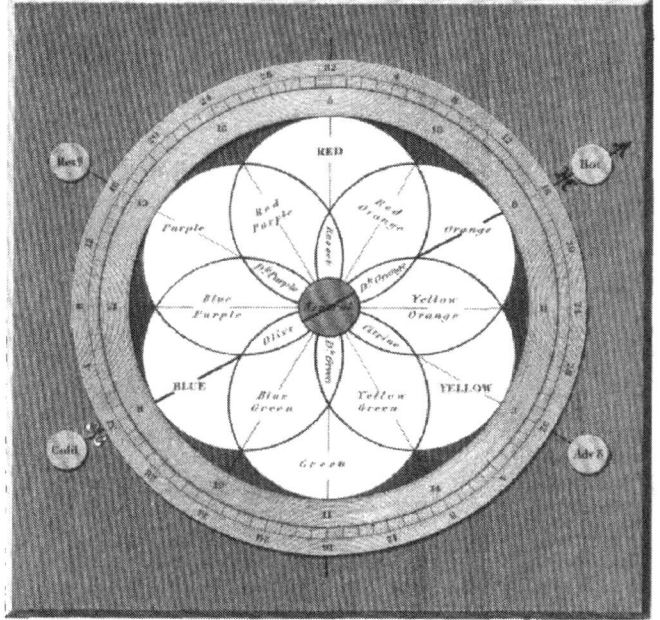

Definitive or Fundamental Scale of Colours
Fig. 3

CHROMATOGRAPHY;

OR,

A TREATISE

ON

COLOURS AND PIGMENTS,

AND OF THEIR

POWERS IN PAINTING, &c.

BY GEORGE FIELD,

AUTHOR OF
"CHROMATICS; OR, AN ESSAY ON THE ANALOGY AND HARMONY OF COLOURS;"
AND OTHER WORKS.

> " Nostrates juvet artifices, doceatque laborem ; -
> Nec qui Chromaticès nobis, hoc tempore, partes
> Restituat, quales Zeuxis tractaverat olim,
> Hujus quando magâ velut arte æquavit Apellem
> Pictorum archigraphum, meruitque coloribus altam
> Nominis æterni famam, toto orbe sonantem."
>
> DU FRESNOY, v. 257.

LONDON:
CHARLES TILT, FLEET STREET.
SOLD BY BOOKSELLERS, PRINTSELLERS, AND ARTISTS' COLOURMEN.
1835.

TO

SIR MARTIN ARCHER SHEE,

PRESIDENT OF THE ROYAL ACADEMY,

AND TO

THE ARTISTS OF BRITAIN.

GENTLEMEN,

THE subject of the following Treatise is so essentially your own, that this address would be vain and uncalled-for, were it not due as a mark of respect and grateful attachment to yourselves, and an acknowledgment of the constant approbation and friendly attention with which you have encouraged the author, and identified him with yourselves and your pursuits.

You, Gentlemen, have attained imperishable honour, by achieving for your country the only requisite to her transcendent reputation—PREEMINENCE IN ART; and it needs not inspiration to foretell, that, by engrafting a refined taste on the prescripts of nature and science, you will consummate a school of colouring which is already celebrated and followed throughout Europe; and as the Greeks of old gave to succeeding ages models of perfect form, so you will bequeath to posterity standards of perfection in colour.

To become humbly instrumental to your progress by the improvement of your palette, is the design of this work, and the constant study of,

    Gentlemen,

      Your much obliged

        and faithful Servant,

          THE AUTHOR.

Cottage, Syon Hill-Park, 1834.

NAMES OF THE SUBSCRIBERS.

A.

Abraham, Henry R., Esq., John Street, St. James's Square
Ackermann, Mr. Rudolph, Regent Street, 3 copies
Aders, Mrs., Savage Gardens, Trinity Square
Ainsley, Mr. Samuel James
Allnutt, Zachary, Esq., Henley, Oxon
Ambrose, Charles, Esq., Newman Street
André, James P., Jun., Esq., Colebrook Terrace, Islington
Applegath, Augustus, Esq., Crayford, Kent
Arnald, George, Esq., A.R.A., Weston Street, Pentonville
Arzoné, Mr. J., Dean Street, Soho

B.

Beechey, Sir William, R.A., Principal Painter to the King, Harley Street
Baker, Mrs. Frances, Asylum, Lambeth
Ballard, Mr., Bookseller, Maiden Lane, 3 copies
Ballingall, Joseph, Esq., Bermudas
Barker, Benjamin, Esq., Smallcombe Villa, Bath
Bartholomew, Valentine, Esq., Foley Place
Beale, James, Esq., Cork
Bennet, Mr., Derby
Berger, Capel B., Esq., Hackney
Blunt, Charles J., Esq.
Bond, John, Esq., Newman Street
Bond, William, Esq.
Boxall, W., Esq., Rome
Brett, T. W., Esq., New York
Briggs, Mr.
Brockedon, William, Esq., Devonshire Street, Queen Square

Brown, Mrs., Liverpool
Brown, Miss Mary Ann, Ditto
Bulwer, The Rev. James
Burnet, John, Esq., Whitehead's Grove, Chelsea
Burnett, Professor, King's College, F.L.S.

C.

Callcott, Augustus Wall, Esq., R.A., Kensington Gravel Pits
Callcott, Mrs. A. W., Ditto
Calrow, Joseph, Esq., St. Mary-at-Hill
Campbell, James Drummond, Esq., of Margate
Carpenter, Mrs. W., Nottingham Terrace
Carpenter and Co., Messrs., Old Bond Street, 6 copies
Cerf, Henry, Jun., Esq , Brussels
Chalon, John James, Esq., A.R.A., Great Marlborough Street
Chambers, George, Esq., Park Village
Chinn, Samuel, Esq., Wilmot Street, Derby
Clover, Joseph, Esq., Newman Street
Clutterbuck, Thomas, Esq., F.S.A.
Collins, William, Esq., R.A., Porchester Terrace, Bayswater
Constable, John, R.A., Charlotte Street, Fitzroy Square
Cooke, William B., Esq., Charlotte Street, Bedford Square
Cooper, Abraham, Esq., R.A., New Milman Street, Foundling Hospital
Cope, C. W., Esq., Rome
Cowen, William, Esq., Newman Street
Cregan, Martin, Esq., P.R.H.A., Dublin
Creighton, John, Esq., Lavenham, Suffolk
Creswick, Thomas, Esq., Hatfield, Herts
Cruikshanks, F., Esq., Great Marlborough Street
Curtis, William, Esq.

SUBSCRIBERS' NAMES.

D.

Dagley, Richard, Esq., Earl's Court, Kensington
Dahmen, John, Esq., Meadow Cottage, Feltham, Middlesex
Daker, Francis, Esq., Whitecross Street
Daniel, Mr. A. K., Maple Grove, State of Ohio
Davis, William Henry, Esq.
Davy, Messrs. Robert, and Son, Newman Street
Day, William, Esq., Isleworth
De la Fons, John Palmer, Regent's Park
Denley, Mr. J., Jun., Old Compton Street, Soho Square, 2 copies
Dingley, Miss, Highgate, Middlesex
Dixcee, Mr. Thomas, Gloster Place, Chelsea
Doo, George T., Esq., Camden Town
Durham, Cornelius, Esq., Arundel Street
Durlasher, Lewis, Esq., Old Burlington Street

E.

Eastlake, Charles Lock, Esq., R.A., Upper Fitzroy Street, Fitzroy Square
Edmonstone, R., Esq., Greek Street, Soho Square
Etty, William, Esq., R.A., Buckingham Street
Eyre, James, Esq., Derby

F.

Froude, John Spedding, Esq., Lincoln's Inn Fields
Field, Miss Catherine W., Hurley House, Berks
Field, William, Esq., Temple, Berks
Fletcher, Joseph, Esq., Chiswick, Middlesex
Foggo, James and George, Esqrs., Warren Street, Fitzroy Square
Frearson, John, Esq., Royal Academy
Frost, William Edward, Esq.
Fuller, Messrs. S. and J., Rathbone Place

G.

Gee, Miss, Isleworth, Middlesex
Glossop, The Rev. Henry, M.A., Corp. Chr. Coll., Cambridge, Vicarage, Isleworth
Green, Mr., Bookseller, Castle Green, Bristol

H.

Hall, James, Esq.
Harding, J. D., Esq., Gordon Square
Hargreaves, Thomas, Esq., Liverpool
Harrison, Cornelius, Esq., Naples
Hart, S. A., Esq., Gerrard Street, Soho Square
Heaphy, T., Esq., St. John's Wood Road
Hewlett, James, Esq., Isleworth
Hofland, T. C., Esq., Pembroke Square, Kensington
Howel, Samuel, Esq., Foley Street
Hughes, Miss Louisa, York Street, Gloster Place
Hughes, Mr. W., Wharton Street, Pentonville

I.

Ireland, Mrs., Kew Road, Richmond

J.

Jennings, W. George, Esq.
Jewell, Joseph, Esq., Clock House, Ashford, Middlesex
Johns, A. B., Esq., Plymouth
Johnson, Henry, Esq., Balham, Surry
Jones, George, Esq., R.A., Duke Street
Joseph, George Francis, Esq., A.R.A., Charlotte Street, Fitzroy Square
Joseph, Samuel, Esq., Charlotte Street, Fitzroy Square
Joseph, Miss Priscilla

K.

Knowles, Sir Francis C., Bart., M.A. F.R.S. F.R.A.S., Park Street, Grosvenor Square
King, John, Esq., Soho Square

L.

Lawrence, Sir Thomas, late P. R. A.
Lamb, Miss, of Bolton, Lancashire
Lance, George, Esq., Upper Seymour Street, Euston Square
Landmann, Colonel, United Service Club
Landseer, Edwin, Esq., R. A., St. John's Wood

SUBSCRIBERS' NAMES.

Lane, S., Esq., Greek Street, Soho Square
Law, Mr., Charlotte Street, Blackfriars' Road
Leslie, Charles Robert, Esq., R. A., Edgeware Road
Lincoln, John, Esq., South Street, Grosvenor Square
Longman, Rees, and Co., Messrs., Paternoster Row
Lovatt, Mr., Green Heys, Manchester
Lynch, Daniel, Esq., Bath Place, New Road

M.

Mulready, Wm., Esq., R. A., Linden Grove, Bayswater
Macpherson, Melville, Esq.
Marsden, Richard, Esq.
Martin, John, Esq., Alsop's Terrace, Paddington
Mosses, Alexander, Esq., Liverpool
Muller, Martin Charles, Esq., Bermudas
Munsey, The Rev. William, M. A., Hereford

N.

Northumberland, His Grace the Duke of, K.G., F. R. S., F. S. A., F. L. S.   2 copies
Northumberland, Her Grace the Duchess of
Needham, Colonel
Nesfield, W. A., Esq.
Newberry, Mr. George John
Newman, Mr. James, Soho Square,   12 copies
Nicolay, Mrs. General, Upper Berkeley Street

O.

Ogle, Lady Letitia

P.

Peat, The Rev. Sir Robert, D.D., Parsonage, Brentford
Parker, Samuel, Esq., Argyle Place
Parris, E. T., Esq., Historical Painter to the Queen, Grafton Street, Bond Street
Peyton, Richard, Esq., Cook's Court, Lincoln's Inn
Phelps, P., Esq., Gt. Marlborough Street
Phillips, Thomas, Esq., R. A., George Street, Hanover Square

Potter, Henry, Esq.
Preston, Thomas, Esq., R. N., Norwich
Purton, William, Esq., Hampstead
Pyne, J. B., Esq., Bristol

R.

Rainier, Peter, Esq., Albany
Reeves and Sons, Messrs., Cheapside, 2 copies
Reinagle, Ramsay Richard, Esq., R. A., Fitzroy Square
Renton, J., Esq., West Street, Finsbury Circus
Richmond, George, Esq., Beaumont Street
Richter, Henry, Esq., Cadogan Place
Rickets, F., Esq., Northcote, Westbury, Gloucestershire
Roberson and Miller, Messrs., Long Acre,   30 copies
Robertson, Andrew, Esq., Berners Street
Robertson, Charles John, Esq., Worton House, Middlesex
Rogers, W. P., Esq., Regent's Park, 2 copies
Rolls, Thomas, Esq., Marlow
Roods, Thomas, Esq., Newman Street
Rothwell, Richard, Esq., R. H. A., Palozzo Romatorke, Rome
Rowbotham, Thomas, Esq., Bristol
Rowney and Co., Messrs. George, Rathbone Place,   2 copies
Rowe, P., Esq., Norton Street, Portland Place

S.

Shee, Sir Martin Archer, P. R. A.
Strange, Sir Thos., Harrow-on-the-Hill, Middlesex
Strange, Lady, Ditto
Salaman, Miss
Sass, Henry, Esq., Charlotte Street, Bloomsbury Square
Say, F. R., Esq., Weymouth Street
Sellers, Leonard Bezer, Esq., Kensall Green, Middlesex
Severn, Joseph, Esq., Rome,   2 copies
Sims, George, Esq., Kensington

Simson, George, Esq., S. A., Edinburgh
Sintzenich, Gustav., Esq., John Street, Fitzroy Square
Smith, Charles, Esq.
Smith, Mr. Henry
Smith, Caterton, Esq.
Smith, Thomas, Esq., Dover Road
Smith, Mr. Charles, Mary le Bone Street, Regent's Quadrant, 2 copies
Solly, Richard Horsman, Esq., F. R. S., F. S. A., F. H. S., M. R. I. A., &c.
Stanfield, Clarkson, Esq., A. R. A., Mornington Crescent
Stothard, Thomas, Esq., R. A.
Stratford, Wm. Samuel, Lieut. R. N., F. R. S., F. R. A. S., Superintendent of the Nautical Almanac
Stump, S. J., Esq., Cork Street
Sutherland, Miss, Grange Walk, Surrey
Syme, John, Esq., S. A., Edinburgh

T.

Trotter, Sir Coutts, Bart., Grosvenor Square
Tabart, George, Esq.
Taylor, Mr. Henry, Charlotte Street, Bloomsbury
Thorn, Miss S. E.
Thwaites, Major, National Gallery
Tomkins, Charles, Esq., Walcot Place
Townend, William, Esq., Liverpool
Trimmer, The Rev. Henry Scott, Vicarage, Heston, Middlesex
Turneau, John, Esq., Liverpool
Turner, Joseph Mallord William, Esq., R. A., Sandicombe Lodge, Richmond
Turrell, Edmond, Esq., Clarendon Street, Clarendon Square

U.

Ubsdell, R. H. C., Esq., Portsea, Hants
Uwins, Thomas, Esq., A. R. A., Charlotte Street, Fitzroy Square

V.

Valpy, A. J., Esq., M. A.
Varley, Cornelius, Mr., Charles Street, Clarendon Square

W.

Williams, Thomas Peers, Esq., M. P., Temple House, Berks
Wright, Thomas, Esq., Upton Hall, Newark, Notts, 2 copies
Wright, Mrs. T., Do.
Wallace, Mrs., Smallcombe Villa, Bath
Walter, George, Esq., East Dulwich
Ward, James, Esq., R. A., Roundcroft Cottage, Cheshunt, Herts
Ward, William Henry, Esq., Brixton Hill
Warren, Ambrose W., Eaton Street, Pimlico
Watts, Walter Henry, Esq., Hampstead Road
Weale, Mr., Architectural Library, Holborn
Westall, Richard, Esq., R. A., Russell Place, Fitzroy Square
Wilkie, David, Esq., R. A., S. A., Kensington
Wilkin, Francis W., Esq., Newman Street
Windsor, J. W., Esq., Bath
Williams, Mr. J., Library of the Fine Arts, Charles Street, Soho Square 2 copies
Winn, Miss, of Nostell Priory, Yorkshire
Winsor, Mr. William, Rathbone Place
Wirgman, Thomas, Esq., Timberham Lodge, Surrey
Withers, Edward R., Esq.

PREFACE.

THE progress of the Art of Painting under the happy auspices of this favoured country, the refinement of taste which it has so universally diffused, and the predilection which prevails for its study and practice as a necessary branch of polite education, render acceptable whatever can facilitate the acquisition, or advance the ends, of this useful, elegant, and enlightening accomplishment. Nor are the concerns of this art uninteresting in a still higher view, since whatever refines the taste, enhances the powers and improves the disposition and morals of a people,—and whatever improves the morals, promotes the happiness of man, individual and social. Hence the high moral and political value of this art, to say nothing of its commercial and religious uses, upon which so much stress has been justly laid.

Among the means essential to proficiency in Painting, none is more important than a just knowledge of Colours and Pigments—their qualities, powers, and effects; and there is none to which the press has hitherto afforded fewer helps. There have appeared, it is true, at different times, several works professing this object, and most of our encyclopædias and

PREFACE.

books of painting treat cursorily on this branch of the art; but not only are these for the most part transcripts of the same obsolete originals, unsuited to the present state of the art, but they are inadequate, irrelevant, and often erroneous or untrue, as every one acquainted with the subject is aware. Hence have arisen several inducements of the author to attempt a guide to the knowledge of colours and pigments generally, and with reference to the Art of Painting in particular.

Most technical readers are fond of recipes and devices communicated as secrets of art, which are accordingly liberally supplied by the caterers to this taste, who compile them in general upon very vague authority; hence the author anticipates some dissatisfaction from those who are in search of royal roads to knowledge, or stratagems and secrets of art: but one principle is worth a hundred processes,—nor was it by prescription, but by the spirit of philosophy, that the Greeks carried the arts to sublime perfection. Hence it is not a detail of the processes for producing pigments that is here intended, which belongs to another extensive art not to be learnt from brief recipes, and upon which the author has a distinct work in hand; * and it is a pursuit accidental and subordinate to painting, in which the pictorial artist can never attain the skill of the chemical colourist without a proportionate sacrifice of his own art, if not, unhappily, of his fortune also,—as was the case with Parmegiano, and has been with others in our own day, who

> Their time in curious search of colours lose,
> Which, when they find, they want the skill to use!
>
> SHEE.

However imperative such sacrifice might have been to the earlier pain-

* Which will complete his original intention, expressed in his "Chromatics," of treating on the *relations*, the *nature*, and the *preparation* of colours, &c.

ters, there is no want in the present day of furniture for the palette,—since pigments, and fine ones too, so abound, that nearly as much experience is requisite to a judicious selection of them as was formerly required for their acquisition or production; to which also there is little temptation, since the expence of the palette, which was immense to the antient masters, is comparatively trifling to our contemporaries. The principal object of the present Treatise is, therefore, by pointing out the true character and powers of colours and pigments, to enable the student to choose and employ judiciously those which are best adapted to his purpose, and thereby to prevent the too frequent disappointment of his hopes and endeavours by a failure at the very foundation of his work. Such failures are often attributed to bad materials; but whatever practices may have formerly prevailed for imposing false and adulterate articles upon the artist, either through ignorance or fraud, it is due to the respectable colourmen of the present day to bear testimony to the laudable anxiety and emulation with which they purvey, regardless of necessary expence, the choicest and most perfect materials for the painter's use; so that the odium of employing bad articles attaches to the artist, if he resort to vicious sources or employ his means improperly. As, however, perfection in all art is a vanishing point, there will be always something to desire in colours, vehicles, and in all the materials of painting; and since the necessity for, and the practice of, the artist in preparing his own materials has ceased, it is the more essential that he should be enabled by precept to select, appreciate, and understand the pigments and colours he employs.

As colours or pigments* refer to the various modes in which painting is practised, and as these modes differ most essentially in the mechanical

---

* The term colour being used synonymously for pigment, is the cause of much ambiguity, particularly when speaking of colours as sensible or in the abstract; it would be well therefore if the term *pigment* were alone used to denote the *material colours* of the palette.

application of colours, in their chemical combinations, and in the purposes to which they are applied,—the chemical and mechanical properties of pigments have been indicated herein, and the appropriate application of each pointed out, so far as to enable the student in each mode to make his own selection; and, with a view to the same end, Lists or Tables of Reference are subjoined, in which pigments are classed according to their uses, properties, and propensities.

A due selection and employment of colours materially is not alone sufficient,—an adequate knowledge of their reciprocal, sensible, and moral influences in painting, is essential to the production of their full effects on the eye and the mind; and, notwithstanding these effects and influences, belong to the higher aims of the colourist, and are of a theoretical bearing, the subject is so connected with the primary object of the work, that it forms also a feature thereof, in subordination nevertheless to practice;—for colouring, like every other art that has its foundation in nature, refers to a whole, and cannot be rightly comprehended, nor perfectly practised, without some attention to all its parts;—hence also the physical causes, relations, and expression of colours have been briefly investigated therein.

To those who choose to study colours philosophically, or to amuse themselves in the ample field of the colourist, even independently of the art of painting, some details of the author's experience have been appended, interesting for their own sake, and not without reference to the cultivated mind of the painter, who exercises his art with an intelligence beyond mere instinct and imitation. Nor is this department of his work devoted to mere rational amusement or mental satisfaction, but aims at fixing some of the principles of colouring upon the ground of science,—at establishing a metral *standard of colours*, which may be of general practical utility; and contributing toward a new and improved theory of vision, light, and colours.

So much for the design of this performance, which might have been augmented with much additional matter, had the limits of the work permitted: some few things not included in the above, but in useful connexion therewith, have, however, been touched on, for which the reader is referred to the Index. As to the peculiar form this attempt has taken, it is to be attributed to the request of an eminent publisher of works of art, that the author should render a subject, which might be dry to many readers, more popular than scientific: he may, nevertheless, have failed in this particular, since he is unconscious of any talent for popularity. Whether or not the author will, upon the whole, have succeeded in the accomplishment of a useful purpose, he professes his intention to have done so, and the foundation of his attempt throughout upon truth, actual observation, and experiment,—principally in the view of the artist,—partly in that of the chemist and natural philosopher, and divested, as much as might well be, of the technicalities which keep these arts asunder.

Should the artist, as he will, find herein matters of his previous knowledge and observation, he will reflect that every reader has not the skill and experience of an artist; and if he meet with things erroneous, the author courts correction and improvement; while in return he tenders his own experience to the inquirer in any way connected with the art. With respect to the application of colours in painting, recourse must be had to practice under the direction of an able master, several of whom have published valuable works of instruction in the various branches of the art *—for of this the student may be assured, that, however useful recipes may be in cookery and pharmacy, the skill of colouring is not

* Such are Dagley's Compendium of the Theory and Practice of Painting, in which the elements of the art are treated with classical simplicity and method;—Mr. Harding's ingenious and admirable Treatise on the Use of the Black Lead Pencil, &c.; Mr. Burnett's elegant performances on Composition, Chiaroscuro, and Colouring; and various others on different departments of the art.

to be acquired by any such off-hand processes ; and that perfect success therein will require—what no literary work can supply—the constant and united efforts of an able hand, a good eye, and a cultivated judgment—directed in the first instance to the works of good colourists, and perfected by an assiduous application to nature.

But though the records of literature and science cannot alone produce a colourist, nor form the practical painter in other respects, they may become most important auxiliaries,—not merely by recreating his faculties, instructing his hand, and extending the sphere of his art, by endless analogies throughout the field of history and philosophy, and the vast regions of poetic fancy ; but also by exciting a just enthusiasm, and stimulating his invention, while they enlarge his judgment and refine his taste ; supporting at the same time that connexion with learning which gives dignity to fine art, and raises it above mere manipulation. Indeed there never was a truly great artist who did not unite with his ability somewhat of literary talent, inclination, or acquirement ; nor is there a surer mark of a low and groveling genius than a contempt of theory or science, and an over-devotion to the mechanical and practical in art; for the connexion of art with science, theory, and practice, and of these with literature, is most intimate and indissoluble; nor is it likely the artist should paint the worse for being acquainted with the philosophy of his art. We derive hence an excuse for having, even in this lowly performance, attempted to draw philosophy on the one hand, and poetry and harmonics on the other, into intimate connexion with colours and colouring, by a variety of natural analogies and poetical instances.

To his Subscribers the author's acknowledgements are due, and his satisfaction will be much augmented if he find in the end that he has not entirely disappointed their expectations. In tendering such acknowledgements he cannot suppress his sentiment of deep regret, that several immediate disciples of the illustrious Reynolds, and other encouragers of this work, have been called from the career of art. Among these are his favourite pupil, the shrewd and amiable Northcote,—the tasteful and admirable Stothard, whose last signature in behalf of art was given to this work with peculiar grace and urbanity,—and the late elegant and courteous President of the Royal Academy, Sir Thomas Lawrence, who was the first to desire his name to be placed on the list of its subscribers.

No praise of the author could do adequate honour to these eminent men: they will receive from posterity the meed of reputation which is due to their various and exalted merits.

CONTENTS.

CHAP. I.
ON COLOURING;—of the Antients and Moderns 1

CHAP. II.
On the Expression of Colour 11

CHAP. III.
On the Relations of Colours 20

CHAP. IV.
On the Physical Causes of Colours, &c. 34

CHAP. V.
On the Durability and Fugacity of Colours 44

CHAP. VI.
On the General Qualities of Pigments 53

CHAP. VII.
On Colours and Pigments individually 58

CHAP. VIII.
On the Neutral, WHITE, and White Pigments 61

CHAP. IX.

OF THE PRIMARY COLOURS.—Of YELLOW, and Yellow Pigments . 72

CHAP. X.

Of RED, and Red Pigments 85

CHAP. XI.

Of BLUE, and Blue Pigments 102

CHAP. XII.

OF THE SECONDARY COLOURS.—Of ORANGE, and Orange Pigments . 115

CHAP. XIII.

Of GREEN, and Green Pigments 121

CHAP. XIV.

Of PURPLE, and Purple Pigments 132

CHAP. XV.

OF THE TERTIARY COLOURS.—Of CITRINE, and Citrine Pigments . 139

CHAP. XVI.

Of RUSSET, and Russet Pigments 144

CHAP. XVII.

Of OLIVE, and Olive Pigments 148

CHAP. XVIII.

OF THE SEMINEUTRAL COLOURS.—Of BROWN, and Brown Pigments . 154

CHAP. XIX.

Of MARRONE, and Marrone Pigments 164

CHAP. XX.

Of GRAY, and Gray Pigments 166

CONTENTS.

CHAP. XXI.
Of the Neutral, BLACK, and Black Pigments 172

CHAP. XXII.
TABLES OF PIGMENTS, indicating their powers, properties, and affections . 183

CHAP. XXIII.
On Vehicles and Varnishes 196

CHAP. XXIV.
On Grounds 212

CHAP. XXV.
On Picture-Cleaning and Restoring 216

CHAP. XXVI.
NEW OPTICAL INSTRUMENTS, and Experiments therewith on Light and Colours 221

Lensic Prism or Prismatic Lens described 222—The Chromascope 223—Primary Colours developed from Shade 224—Primary Colours developed from Light,—their number 225— Primary Colours developed from Colours—their indivisible Triunity 227—Beautiful Prismatic Spectra 228—The Rainbow 233—Metamorphosis of Colours 234—New Ocular Spectra 236—The Eye not cognisant of colours,—phenomena of single vision 236—Colour in the Sensorium 237—Equivalence of Colours 238—Colours of Shadows 239—New Prismatic Spectra 240—Metrochrome or Standard of Colours 244—Simple Colours measured 245—Compound Colours measured 246—Doctrine of Contrasts 247—Tables and Scale of Equivalents 249—Monochrome 252.

NOTES.

On Styles of Painting 255—Colouring 256—Newton's Primaries 257—Semineutral Colours 258—Illustrative Diagrams 259—Elements of Colours 259—Defects of Vision—Eye for Colours 260—Influences of Colours, &c. 261.

CHROMATOGRAPHY;

OR,

A TREATISE ON COLOURS AND PIGMENTS, &c.

CHAPTER I.

ON COLOURING.

" Colouring is the sunshine of the art, that clothes poverty in smiles, and renders the prospect of barrenness itself agreeable, while it heightens the interest and doubles the charms of beauty."
—OPIE's LECT. IV. p. 138.

How early, and to what extent, colouring may have attained the rank of science among the antients, is a question not easily set at rest; but that some progress toward it had obtained, even among the early Egyptians, is a fact proved by the late researches of the Messrs. Salt, Beechey, and Belzoni, who have again opened to us the magnificent tombs of the Egyptian kings at Thebes. The former of these gentlemen has described the walls of the royal mausoleum as covered with paintings in fresco, so fresh and perfect as to require neither restoration nor improvement: so far from it, that neither of those artists, with all their talents and attention in copying them, found it possible to equal the brilliancy of the originals; which, according to the literal expression of Mr. Salt, " as far as colours go, throw all others completely in the back-ground :" he adds, " the most minute attention and painful labour are not equal to give a faithful idea of the fascinating objects of these designs. The scale of colours in which they are painted is that of using pure vermilion, ochres, and indigo; and yet they are not gaudy, owing to the judicious balance of the colours and the artful management of the black. It is quite obvious that they are worked on a regular system, which had for its basis, as Mr. West would say, the colours of the rainbow, as there is not an ornament throughout the dresses where the red, yellow, and blue are not alternately mingled, which produces a harmony that in some of the designs is really delicious."

It may however be remarked, that as these paintings were viewed by torch-light, it is probable the blue and yellows, which looked like indigo and ochres, suffered in appearance from the colour of the light, and were perhaps painted with a yellow brighter than ochre, and with the celebrated Armenian blue, of a livelier character than indigo; and, if not identical with our ultra-marine, was even of greater pecuniary value and esteem among the antients: and, with respect to the harmony of these paintings, it is obvious that it was of the simplest kind,—a first step that extended only to the crude accordances of the primary colours, and precisely that of the native Mexicans, South Sea islanders, and the North American Indians at this day.* So late indeed as Van Eyck, and the earlier masters of the German school, the practice of colouring had in it much of the same primeval character.†

The above account of the ancient Egyptian painting is fully borne out by the testimony formerly given by Diodorus Siculus, Norden, Dr. Perry, and, recently by M. Champollion and others. Norden remarks, " that the manner of painting is so totally different from any thing in practice at this time, as to make it necessary for me to give you some slight idea of it. A painting eighty feet high, and proportionably broad, is divided into two ranges of gigantic figures in bas-relief, and covered with most *exquisite colours*, suited to the drapery and naked parts of the figures. But what is still more wonderful is this—that the *azure, the yellow, the green, and the other colours made use of, are as well preserved as if they had been laid on but yester-*

---

* Since writing the above, we have been favoured with some beautiful Indian work from Nova Scotia by the lamented lady of the noble admiral, late chief on the North American station, in which the combinations of the primary colours, and management of black and white, are perfectly illustrative of the remarks of Salt, Norden, and others, respecting the colouring of the ancient Egyptians.

† Examples of which may be seen in the admirable collection of Charles Aders, Esq, to whom, and to whose talented lady, the author is indebted for frequent opportunities of confirming this observation.

day, and so strongly fixed to the stone that I was never able to separate them in the least degree."—*Templeman's Norden's Travels*, p. 33.

Dr. Perry, who visited Upper Egypt about the beginning of the last century, describes, among the stupendous ruins at Carnac, on the site of antient Thebes, an apartment " one hundred paces wide, and sixty deep; perfectly crowded with pillars twelve feet in diameter, and seventy-two high: all these columns, as well as the ceiling, roof, and walls of the apartment, are quite covered or crowded with figures in basso-relievo and hieroglyphics; all exquisitely beautiful, and finely painted all over; and, which may seem very extraordinary, all these things look as fresh, splendid, and glorious, after so many ages, as if they were but just finished."—*Perry's View of the Levant*, p. 341.

M. Champollion, jun. has given a similar account; and all agree in describing these colourings of the Egyptians in the most glowing terms of admiration.

From Philocles, the Egyptian, we have historical evidence that the Greeks obtained the seed of their *Ars Chromatica*, which the latter are said to have carried by gradual advances of the art during several centuries from the *monochromatic* of their earlier painters, to the perfection of colouring under Zeuxis and Apelles. Thus the principles of light, shade, and colours in painting appear to have been understood by the antient Greeks to have been lost with their valuable treatises on the art, including that of " Euphranor on Colours," ever since the time of the Romans, and not to have been recovered at the restoration of learning in Europe. Accordingly M. Angelo, Raffael, and all the early Roman and Florentine painters, so eminent in other respects, were almost destitute of those principles, and of all truly refined feeling of the effects of colouring.

The partial restoration of this branch of the art of Painting, if not even its invention, seems to have been coeval with oil-painting; and the glory of it belongs to the Venetians, to whom the art passed with the last remains of the Grecian schools and their productions, after the capture of Constantinople at the beginning of the thirteenth century, and among whom Giovanni Bellini laid its foundation, and Titian carried it to its highest perfection. From the Venetian it passed to the Lombard, Flemish, and Spanish schools. It is to be doubted, notwithstanding, whether there was not as much of instinct as principle in the practice of these schools, and that colouring remains yet to be established in its perfection as a science.

The *historical distribution* of Painting according to the schools is not perhaps exactly coincident with its true, natural, and *philosophical classification*, according to which there are but three principal classes or schools; viz. the gross and *material*, which aims at mere nature, to which belong the Dutch and Flemish schools; the *sensible*, which aims at refined and select nature, which accords with the Venetian school; and the *intellectual*, which corresponds with the Greek, Roman, and Florentine schools, and aims at the ideal in beauty, grandeur, and sublimity: and it is somewhat remarkable, in a scientific view, that these schools should have retrograded.

If the excellence of the Roman and Florentine schools in the high departments of figure, composition, and expression, must be admitted, they fail, nevertheless, in the just effect of an art which addresses itself to the mind through the sight. Their works, accordingly, have often as little effect upon the *eye*, as the finest poetry badly set to music has upon the *ear;* and as this would be better without the music, so those would often be better without their colour. True, natural, and unaffected taste, which admits no unresolved discordance among its objects, will therefore prefer generally the Venetian, to the Roman and Florentine schools, because it excels in that which is the essential basis of the art and its end of pleasing, by the medium of sensible impression.

Upon the same principle, the sublimest sentiments delivered, however accurately, in language unmeasured and inharmonious, will never redeem the performance of the poet, nor raise it above more ordinary thoughts delivered in the true measure and melody of speech; for these are the first essential,—the constituent matter,—the very *colour* of the poet's art.

> Poets are Painters;
> Words are their paint by which their thoughts are shown,
> And Nature is their object.
>
> GRANVILLE.

Or, as Horace thus briefly expresses it:—
> "Ut Pictura Poesis erit."

So also, according to a correct analogy, colouring may be called the eloquence of painting,—the animating principle which gives life and action to the fine thoughts of the painter.

"Among the several kinds of beauty," says Addison, "the eye takes most delight in colour. We no where meet a more glorious or pleasing show in nature, than what appears in the heavens at the rising and setting of the sun, which is wholly made up of those different stains of light, that show themselves in clouds of a different situation. For this reason we find the Poets, who are always addressing themselves to the imagination, borrowing more of their epithets from colour, than from any other topic."—If then the purpose of painting be analogous to that of poetry, how much more powerful and important must the expression of colour be to the painter than to the poet; and how absurd the affectation of the artist or critic who undervalues colouring,—the sole object of sight,—the sole matter of Painting, whose mistress is Nature, and to her only we need appeal for evidence of its powerful effects in grandeur, as well as in sublimity and beauty.

In the practice of the individual in Painting, as well as in all the revolutions of pictorial art, in ancient Greece as in modern Italy, colouring has been the last attainment of excellence in every school;* thus Zeuxis succeeded and excelled Apelles in colouring, as Titian did Raffael. There is hence just reason to hope the artists of Britain will transcend all preceding schools in the chromatic department of Painting, if even in their progress they should not surpass them in all other departments, and in every mode and application of the art, as they have already done in an original and unrivalled use of water-colours in particular, in the perfection of landscape, in the new and beautiful device of panoramic perspective, and in engraving.

Happily too, a school of colouring has arisen, that confirms this expectation, strengthened also by the suitableness of our climate to perfect vision, —by that mean degree of light which is best adapted to the distinguishing of colours,—by that boundless diversity of hue in nature, relieved by those fine effects of light and shade which are denied to more vertical suns,—and by those beauties of complexion and feature in our females, by which we

---

* And that also of rarest occurrence, if not of greatest difficulty;—hence Du Pile justly remarks, "that for near three hundred years since Painting was revived, we could hardly reckon six Painters that had been good colourists." He might have added—" among thousands who had laboured to become such."—DU PILE'S DIALOGUE, p. 28.

are perpetually surrounded;—respects in which at least our country is not unfavourable to art. In many obvious references too, this country resembles Venice of old, the greatest of whose antient glories is still her Titian, her Giorgione, and her school of colouring; and England has had her Reynolds, and Wilson, and still has living colourists, of whom we will not offend the modesty, nor distinguish invidiously. It has, however, been urged to the disparagement of the British school, that it excels in colouring; as if it were incompatible with any other excellence, or as if nature, the great prototype of art, ever dispensed with it.

This appeal from the decisions of criticism,* in behalf of colouring, is not intended to militate against the necessity the Painter is under, of studying the other branches of his art, nor to assert the redeeming power, or the exclusive excellence, of colouring.†

> For 'tis the MIND that makes the body rich;
> And as the sun breaks through the darkest clouds,
> So Honour 'peareth in the meanest habit.
> What! is the jay more precious than the lark
> Because his feathers are more beautiful?
> Or is the adder better than the eel,
> *Because the painted skin contents the eye?*
>   SHAKSP., TAMING OF THE SHREW.

Colouring alone will not therefore constitute a picture; still colour is the flesh and blood of the art,—and if it be wanting, the finest performances will remain lifeless skeletons, and fail to *please;* and as the proper end of painting is *to please,* and there is a higher and more effectual medium for addressing mind, the most intellectual performances of the painter, and the grandest efforts of his invention, will fall short of their true purpose, if they pass not to the mind by the medium of pleasurable effect through their appropriate sense of sight, of which colour—and colour alone—is the immediate object; and what is Painting altogether but the art of representing visible things by light, shade, and colours? Colouring is therefore the first requisite, the matter and medium of the Painter's art: it is indeed

---

* See Note A.

† As music relates to *sound* simply, and poetry, or *figurative speech,* to signification; and as these when united become *sound significant,* so it is with *colouring* in respect to *Figure, &c.;* the first belongs principally to the harmony of Painting, the latter to its sentiment or poetry, while in the perfect picture they are united.

the first quality which engages attention and regard—the best introduction to a picture, and that which continues to give it value so long as it is regarded. In the grosser matters of taste a food or medicine may be both salutary and nutritious, but we nauseate it if it be not also palatable or *well-tasted:* such is Painting without colouring, and so it is with all objects of sense; nor did the first and greatest critic that ever lived assign any higher end than *pleasure* to even Poetry itself.* It was the deficiency of colouring in the great works of the Roman and Florentine schools which occasioned Sir Joshua Reynolds, with such admirable candour, to confess a want of attraction in their works, and to declare the necessity of a forced and often-repeated attention, with previous cultivation and profound investigation of their other excellences, to a just relish and estimation of their greatness, which hundreds have affected to admire upon authority, without feeling or comprehending.† For this deficiency in the colouring of these great masters, apologies more ingenious than just have been offered by eminent critics, to the perversion of taste and truth; while some, through false admiration or want of *sense*, have attributed fine colouring to these masters.

It is the consecration of great names which blinds, betrays, and ruins their followers; and it is no less true than lamentable, that the modern Italian schools have fallen a sacrifice to the greatness of their models. And so it is in all sciences when great human authorities have subverted the authority of nature—the master of masters!

> Nature is made better by no mean,
> But Nature makes that mean. So o'er that art
> Which you say adds to Nature, is an art
> That Nature makes. You see, sweet maid, we marry
> A gentler scion to the wildest stock,
> And make conceive a bark of baser kind
> By bud of nobler race. This is an art
> Which does mend Nature—change it rather,—but
> *The art itself is Nature.*
> SHAKSP., WINTER'S TALE.

With respect to those departments of Painting which have been ranked above, and represented as inconsistent with colouring, it may be questioned whether this is not to be attributed to a proneness, common enough in all cases, to consider the *greatest difficulties* as the *highest attainments* of art;

---
* Aristotle's Poetics.   † See Note B.

that which is *most rare* as of *highest esteem*, and to the mistaking of *novelty or singularity* for *beauty and excellence ;*—but we have to remember that that which is *most beautiful*, like that which is most useful, is least rare in nature ; nay, it is beautiful abstractly, because *it is not rare*. Thus the most common form of any thing is the most general or middle form, and such the Greeks have taught us is also the most beautiful and natural. We have to consider also, that colour individually gives finish or final value to all the productions of nature, not excepting the diamond ;—and that he who has attained colouring in its complex and higher relations, has rivalled Nature in her chief beauty, whatever we may determine it to be in art.

Indeed the greatest masters of design in every school have given ready testimony to the claims of colouring; and, according to Vasari, even Michael Angelo himself, the greatest of them all, conceded to Titian, on viewing his Danaë, that he wanted nothing but the correctness of the Roman school to have rendered him the greatest painter that ever lived. There never was, indeed, is not, nor ever will be, any painter who does not colour well, for any other reason than because he is not able; and he, who should teach neglect of colouring as a doctrine, will find no disciple of true feeling and ability who would follow him. No master individually, nor school collectively, can be regarded as perfect without this first and last accomplishment of the art; nor did Painting attain its immortal reputation in Greece until Zeuxis and Apelles had conjoined colouring to the climax of its excellences, nor will such high repute ever attach itself to any modern school without it.

Every one knows the false taste and absurdity which have sprung from the admiration of difficulty and novelty, in place of the natural and expressive, in the sister art of music ; and it behoves the true lovers of art to guard against similar degeneracy in Painting, that she may not sink in like manner by quitting the charms of nature for those of artifice and false refinement, nor even abstract herself in those sublimities or excellences exclusively, which the artist alone can appreciate.

Considering that the evidence of the eye is superior to that of the ear, and that the science of colours should be naturally easier than that of sounds, it is remarkable that music should have taken the advance of other sciences, and that colouring, as a science, remains yet in the rear. This precedence the former may perhaps owe to its more sensual character, as well as to its connexion with poetry:—since, however, these arts are intimately related and analogous, the colourist may justly trust to advance and perfect his science by following wherever the others lead, and particularly so by adopting, as far as possible, the harmonic principles of the musician.

We must not quit our subject without remarking, that there is a vicious extreme in this branch of Painting, and it is that in which colouring is rendered so principal, as by the splendour of its effects on the eye to diminish all other powers of a work upon the mind, or by want of subordination in the general design to overlay the subject;—no excellence of the mere colouring can in this case redeem from censure the performance of the painter. Add to which, there is a negative excellence which belongs to colouring, whence the painter is not always to employ pleasing and harmonious colours, but to take advantage of the powerful effects to be derived from impure hues, or the absence of all colour, as Poussin did in his 'Deluge,' thus well commented on by Opie:—" In this work there appears neither black nor white, neither blue, nor red, nor yellow; the whole mass is, with little variation, of a sombre gray, the true resemblance of a dark and humid atmosphere, by which every object is rendered indistinct and almost colourless. This is both a faithful and a poetical conception of the subject. Nature seems faint, half-dissolved, and verging on annihilation ——." But this want of colour is a merit of colouring, and not its reproach. Vandyke employed it with admirable effect in the background of a Crucifixion, and in his Pieta; and the Phaëton of Giulio Romano is celebrated for a suffusion of smothered red, which powerfully excites the idea of a world on fire, although this artist, like his school, was deficient in the more subtle graces of colouring. In no case, however, is any thing legitimate in art that has not an authority in nature; and in Painting, as in Poetry, *the imaginative* must be founded on *the true*. Without this basis, effect falls into extravagance—grace into affectation—beauty into deformity—and the sublime into the ridiculous. Where truth and nature end, vice and absurdity begin; hence the moral influence of pure art, in which the habits of truth and honesty conduce to success, and are essential in a high degree to all the attainments of genius. The painter may, notwithstanding, deviate from the real into the ideal or abstract, even so far as in some instances to violate probability, but never to transcend possibility. To deviate successfully from objective truth presupposes, nevertheless, both judgment and genius in the painter; that is, the power of justly imagining and generalizing.

To conclude—it is not in painting and decorating,—in the sentiments it excites, nor in the allusions of poetry that the value of colouring is comprised; it has an intrinsic value which, by augmenting the sources of innocent and enlightened pleasure, entitles it to moral esteem. We all know the delight with which music gratifies the ear of the musically inclined.—The lover of art would not for worlds forego the emotion which arises from regarding nature with an artist's eye;—but he who can regard nature with the intelligent eye of the colourist, has a boundless source of never-ceasing gratification, arising from harmonies and accordances which are lost to the untutored eye;—rocks and caves,—every stone he treads on, mineral, vegetal, and animal nature,—the heavens, the sea, and the earth, are full of them; wherever eye can reach or optical powers can conduct, their beauties abound in rule and order, unconfounded by infinite variety; and to assert that colouring permeates and clothes the whole visible universe, incurs no hyperbole.

CHAP. II.

ON THE EXPRESSION OF COLOUR.

"Every passion and affection of the mind has its appropriate *tint*; and colouring, if properly adapted, lends its aid, with powerful effect, in the just discrimination and forcible expression of them; it heightens joy, warms love, inflames anger, deepens sadness, and adds coldness to the cheek of death itself."—OPIE'S LECT. IV. p. 147.

ASSURED as we must be of the importance of colouring as a branch of painting, colours in all their bearings become interesting to the artist. This subject, considered in the whole breadth of its survey, appears to refer to the material principles of colours, to their sensible relations, and to their intellectual effects, or highest purpose.

We will first discuss the latter of these, which is the prime object of colouring, comprehending the effects of colours and colouring on the passions, sentiment, and affections of the mind, and may therefore be termed the *Expression of Colour;* a subject which, if it has not been totally without light, has been involved in much obscurity: and if philosophers have hitherto argued respecting the causes and harmony of colours with little of the confidence of science, they have spoken of their expression and moral effects with more imperfect apprehension, or even with confusion. There may indeed be some who, from natural organic defect or uncultivated sense, will question these latter effects altogether. Yes,

> The man
> Whose eye ne'er open'd on the light of heaven
> Might smile with scorn, while raptured vision tells
> Of the gay-colour'd radiance flushing bright
> O'er all creation ——.
> AKENSIDE.

Yet the enlightened artist acknowledges these effects from having seen and felt them; and having felt them, it becomes a purpose of his art to produce them to the feelings of others gifted by nature or attainment to enjoy them. This alone is sufficient to render these powers of colours the subject of his serious inquiry, and to give value to hints and suggestions which may assist in realizing them in his practice, when, according to the expression of Addison, he is "obliged to put a *virtue* into colours, or to find out a proper dress for a *passion*," &c.—Treatise on Medals, Dial. 1.

For evidence of the natural expression of colours, we need not look beyond the human countenance, that masterpiece of expression, in which are acknowledged—the *redness* indicative of anger and the ardent passions, and the blush of bashfulness and shame betraying a variety of consciousness,—the sallowness or *yellowness* of grief, envy, resentment, and the jealous passions,—the cold, pallid *blueness* of hate, fear, terror, agony, despair, and death; with a thousand other hues and tints accompanied by expressions readily felt, but difficultly described or understood.*

If we turn our view from the face of man to that of nature in the sky, we find colour equally efficient in giving character, sentiment, and expression to the landscape, indicating the calm and the storm, and in infinite ways betraying the latent emotions of the spirit of nature.

It is by this influence that the greenness of spring indicates the youth, vigour, and freshness of the season; that the light, bright, warm, *yellow* hues of summer express its powers; that the glowing *redness* of fruits and foliage denote the richness of autumn,—

> Yon hanging woods that, touch'd by autumn, seem
> As they were blossoming hues of fire and gold.
>   COLERIDGE.

and that *blue*, dark gray, and white tints express the gloom and wintry coldness of nature.

The analogy of the natural series of colours, with the course of the day and the seasons, coincides with the ages of man or the seasons of life, and adapts it to express them in the hues and shades of draperies and effects; from the white or light of the morn or dawn of innocuous infancy,

---

* Whether these colours of the human countenance are to be variously ascribed to the agency of nerves, blood-vessels, or lymphatics;—whether the *warmth and redness* expressing active feeling be not attributable to *arterial action*, and the *cold hues* of passive suffering to *venous re-action;*— and whether the passions denoted by the *sallow or yellow hue* are not *biliary affections*—are questions we leave to the anatomist.

through all the colours, ages, and stages of human life, to the black or dark night of guilt, age, despair, and death.

Throughout all seasons, and in all countries, it is by the colour of his crops that the hopes, fears, acts, and judgments of the husbandman are excited; nor are the colours of the ocean and the sky less indicative to the mariner; nor the colours of his merchandise to the merchant,—so universal is this language of colour, the sole immediate sign to the eye, which is the chief organ of external expression and intelligence.

Whether it be the face of nature or of man that is tinged with the varied expression of the gloomy and the gay, it reciprocates corresponding sentiments in the spectator, and we even form judgments of the disposition, temperament, and intentions, as well as of the youth, vigour, age, and race of individuals by colour and complexion; hence colours have been made symbols of the passions and affections, denoting by a sort of tacit consent their connexion with moral feeling, all of which is transferable to the canvas.

Of these popular symbols, *black* denotes mourning or sorrow; *gray*, fear, &c.; *red* is the colour of joy and love; *blue*, of constancy; *yellow*, of jealousy; *green*, by a physical analogy, of youth and hope; and *white*, by a moral analogy, of innocence and purity.

These remarks do not apply merely to the more positive colours individually, but extend with even greater force to the more neutral or broken compounds, every hue and shade having its corresponding shade of expression and reciprocation, affording materials for the cultivation of feeling and taste; the sublimest expression vibrating in all cases to the most delicate touch. The subject is fertile, but enough has been said to confirm the fact, if it can be disputed, of the general, moral, sentimental, and natural expression of colours, analogous to that of musical sounds; and of the expression of colours individually we shall take further occasion to speak under their distinct heads.

By what mysterious power colours and sounds thus vibrate and reflect these affections, is beyond our present inquiry; if the fact be established, by investigating its instances we may induce or generalogize a theory, or advance our practice, in which we already acknowledge the powers of colours to soothe and delight by gradation of hue and shade, to excite and animate by their various contrasts, and to distract and repel by infraction and discordance.

It may be doubted indeed whether the expression of colour is not naturally more powerful than that of form or figure; for, though form has also its natural expression, it owes its chief force to the auxiliaries of custom, association, and consent; whence lines and forms have almost usurped the office of expression with the painter, but, aided by the influence of colour, they become irresistible. Perhaps colour and form have peculiarities of expression which ought to be distinguished; and, if we might venture an opinion on this head, the expression of form is more powerful in figuring the *passions;* and that of colour, in representing and exciting the more delicate perceptions of internal feeling and sentiment: the one is the rhythmus of expression, and, like those of poetry and music, strikes every eye; the other is the harmony that touches certain natural chords, which vibrate to an eye gifted or cultivated to perceive and feel it.

The choice of colours then which the artist infuses into, and with which he clothes and surrounds his figures, or his scenes and compositions, is by no means arbitrary, nor merely an affair of conformity to the natural object, or of sensible satisfaction to the eye; but has, in its highest view, a rational and moral reference to the mind, dependent on the subject, and the sentiment or moral he means to excite or convey: and our common habits of thinking and speaking coincide herein when we attribute moral and sensible qualities to colours, by denominating them faint or strong, true or false, foul or fair, harmonious or discordant, dead or lively, sedate, fresh, good and bad, modest and meretricious, solemn, gloomy and gay, &c.; and it is hence by tone and colouring that the artist is able to aid and excite the ruling and subordinate sentiments of his performance in the manner of the musician; and that, although he should copy nature in his colouring, he will not do so servilely, but with taste, discrimination, and reference to these ends. There is *the ideal* in colouring, as well as in forms, which belongs to the perfection of beauty and sentiment, which it is the highest office of the painter to attain; and it is that in all these arts to which the philosophic minds of the Greeks aspired: " Is not painting, Parrhasius, a representation of what we see? By the help of canvas and a few colours you can easily set before us hills and caves, light and shade, straight and crooked, rough and plain, and bestow youth and age where and when it best pleaseth you; and when you would give us *perfect beauty* (not being able to find in any one person what answers your idea) you copy from many what is beautiful in each, in order to produce this perfect form."—Xenoph. Mem.

c. x. p. 167. It is the same in colouring, it must be induced or generalized. But of ideal beauty, in every case, nature must supply the means,—not individually, but in the way of selection, generalization, and refinement; for there is no other source of fine ideas in science or in art.

It is in the election of his colours, not less than in their use and arrangement, that the artist merits the reputation of a colourist; and he may perhaps borrow from, as well as contribute something to, the poet, who has not failed to avail himself of the powers of colours on the imagination in exciting, heightening, and extending ideas and sentiments,—in the construction of epithets, the decoration of figures natural and rhetorical, and in all the imagery and witchery of his art. We may indeed truly remark, with respect to the poets, that many of the most exquisite passages of their works are indebted to colours principally for their beauty and effect.

The expression of colour in poetry must of course be limited to the signification of terms, which, with respect to colours, is hitherto confined to their simple names and relations: poetry, therefore, falls far short of nature and painting in this respect; it is nevertheless open to all the refinements of language and art, on which point much remains to be done by the poet, and herein the painter may refund part of the obligation he owes to the bard:

Blend the fair tints, and wake the vocal string.

COLLINS.

Poets, like painters, are comparatively good or bad colourists; and it is remarkable that the poets of nature are invariably the best, while the poets of art, and imitators, are as indifferent colourists as those painters and copyists are who have studied colouring in pictures only. Hence some of the earlier poets, who probably drew their images more immediately from nature, have availed themselves more, and more truly, of the powers of colours than later poets; whence Spenser, and Shakspeare in particular, are painter poets. This remark is not quite applicable to the schools of painting in which, as before observed, colouring has been the latest attainment of the art, although not without exception, nor without traces of natural colouring in early examples of the art.

The latitude and licence of the poet with regard to the use and expression of colour is even wider than that of the painter, and hardly bounded by the usage of nature herself when it suits his sentiment to deviate in this respect; hence with the poet, the sea becomes 'the *black* ocean,' 'the *green* ocean,' 'the *purple* main,' 'the *azure* deep,' 'the *white* waves,' &c.; and so it is with the sky, the land, the forest, or other natural objects. Such also are the coloured garbs in which he clothes the animate figures of gods and goddesses, &c. whereby the various parts of nature, &c. are poetically designated and expressed.

In collating the poets for instances of this *poetical* painting, none appears to our view to have had juster conception of the beauties and powers of colours than our great dramatist, whose genius seems to have been almost universal. Sometimes he harmonizes with the primary colours, as thus—

> Thou shalt not lack
> *The flower that's like thy face, pale primrose;* nor
> *The azured harebell, like thy veins.*
> CYMBELINE.

Sometimes he employs the secondaries, as in the order of Titania to the Fairies to honour her Love,—so much admired by Dryden for its poetic beauty:—

> Feed him with *apricots*, and dewberries,
> With *purple* grapes, *green* figs, and mulberries.—
> And pluck the wings from painted butterflies
> To fan the moon-beams from his sleeping eyes:—
> Nod to him, Elves, and do him curtesies.
> MIDSUM. NIGHT'S DREAM.

In both these instances one of the three colours is kept back, inferred but unexpressed or subdued, as it is generally in nature, particularly in flowers, and even in their species; e. g. we have roses, *red, yellow*, and compounds only, for nature does not produce a *blue rose*, but in its place roses inclined to *purple*, in which blue is subdued by red and black; the same may be observed of the hollyhock and other flowers. Colours nevertheless, it is true, are sometimes given to flowers, &c. in pictures which nature never dared to give; and though the colours may be required in the picture, yet when they are so given, it is an offence to truth, which makes its impression upon the mind of the observer. This adherence to nature and truth—this best policy of honesty in all things, is one of Shakspeare's greatest charms, and belongs to excellence in every intellectual art. How natural, tender, expressive, beautiful, and true, is the following inquiry concerning an occasion of grief:

> What's the matter,
> That this distemper'd messenger of wet,
> The *many-colour'd Iris*, rounds thine eye?

That Shakspeare discriminated nicely in colours is apparent from the following :—

> If you will see a pageant truly play'd,
> Between the *pale complexion* of true love
> And the *red glow of scorn* and proud disdain,
> Go hence a little.

And again :—

> There was a *pretty redness* in his lip;
> A little *riper and more lusty red*
> Than that mix'd in his cheek ; 'twas just the difference
> Betwixt the *constant red* and *mingled damask*.

With what truth and effect he avails himself of the chromatic discord of green and yellow, which he uses metaphorically for freshness and jealousy, by natural feeling or discernment, as if he theorized in colours, in the following hackneyed passage :—

> She never told her love,
> But let concealment, like a worm i' the bud,
> Feed on her *damask cheek :* she pined in thought;
> And, with a *green* and *yellow* melancholy,
> She sat like Patience on a monument,
> Smiling at grief.

The discord therein resolves itself in "damask," which is the perfect contrast or equivalent of "green and yellow." Of this species of contrast in colouring, Shakspeare is a great master; witness the blood of Duncan on the hand of Macbeth, contrasted or opposed by the colour of the ocean :—

> MACBETH. Will all great Neptune's ocean wash this *blood*
> Clean from my hand ? No; this my hand will rather
> The multitudinous seas incarnardine,
> Making the *green* one *red !* *
> LADY M. My hands are of *your colour*, but I shame
> To wear a heart so *white*.

Numberless instances might be adduced of the correctness of his judgment and feeling, in employing the beautiful and peculiar relations and effects of

---

* Making the green 'ocean' red ?

red and white when mingled or opposed; but the latter and following quotations may suffice :—

> I have mark'd
> A thousand *blushing apparitions* start
> Into her face; a thousand *innocent shames*
> In *angel whiteness* bear away those *blushes.*

> Go, prick thy face, and *over-red* thy fear,
> Thou *lily-liver'd boy!*

Not to fatigue by multiplying instances, we refer the inquirer to the sections under which each colour is treated of, for examples more particularly in point from the poets, having preferred here a general illustration from a single authority; of which we have found none equal to Shakspeare, who often produces these chromatic effects by mere allusion, clothing immaterial things in imaginary colours :—

> Thus conscience does make cowards of us all ;
> And thus *the native hue of resolution*
> Is sicklied o'er by *the pale cast of thought!*

He does not deem it necessary to tell us that "the native hue of resolution" is hot and *fiery red*, nor that it is subdued or "sicklied o'er" by the cold, dull, livid "cast of thought;"—the very means the painter would have taken to lower such "native hue." Indeed Shakspeare in the employment of colours always evinces a refined feeling of our art; but in what feeling or sentiment could he be wanting who drew all his resources from the fountains of nature and truth ?—

> And he the Man whom *Nature's* self had made
> To mock herself, and *truth* to imitate
> With kindly counter under mimic shade,—
> Our pleasant WILLY,—ah! is dead of late.
> SPENSER'S TEARS OF THE MUSES.

Milton and other poets abound with fine examples of colouring, but they have not always the natural truth and simplicity of Shakspeare's. Byron's palette is principally set with *black* and *red;* but in this there is something not less characteristic than is the *purple* and *gold* of Homer.

Ere we close this sketch we will subjoin one more illustration, an exception to our intention with regard to the Bard of Avon, from a genuine pupil of nature—a genius of bright and early promise—in this poetical painting,* who, in the following stanzas, has unconsciously, but with just feeling, brought together the entire scale of primary and secondary colours accurately arranged and contrasted, in all the glow of natural imagery:

> 'Twas in a glorious eastern isle,—
>   Where the acacias lightly move
> Their *snowy* wreaths; where *sunbeams* smile
>   Brightly, but scorchingly, like love,—
> Round which the ocean lies so clear,
>   The *deep red coral blushes* through
> The waves that catch its *crimson* hue,
>   While the *soft roseate tints* appear
> Mix'd with *the sky's reflected blue!*
>   Where, brilliant as *the golden rays*
>   That shine when day gives place to night,
>   The shells, that are as *rainbows* bright,
>   Glow through the waters in a blaze
>   Of glorious *gold and purple* light!
> Where *roses* blossom through the year,
> And palms their *green-plumed* branches rear.
>
> MARY ANN BROWN'S ADA.

It may be added, that the female eye seems to be particularly receptive and perceptive of the tender, beautiful, and expressive relations of colours; and we have repeatedly heard it remarked by that graceful painter and colourist, the late lamented President of the Royal Academy, whose subjects were from the high and refined classes of the sex, that in no instance whatever had he occasion to request or desire any change of the colours in which they presented themselves, so judicious and natural was their taste and feeling as to what best suited their peculiarities of character, complexion, and expression.

From the foregoing, it may be concluded that the sentiments effected in the mind by hues, shades, and colours can, next to nature and painting, be nowhere better studied than among the poets; we shall accordingly, as occasion offers, make free use of their productions.

* At sixteen years of age.

CHAP. III.

ON THE RELATIONS OF COLOURS.

"I know not if lessons of colouring have ever been given, notwithstanding it is a part so principal in painting that it has its rules founded on science and reason. Without such study it is impossible that youth can acquire a good taste in colouring, or understand harmony."
MENGS ON THE ACADEMY AT MADRID.

"He that would excel in colouring must study it in several points of view; in respect to the whole and in respect to the parts of a picture, in respect to mind and in respect to body, and in regard to itself alone."—OPIE'S LECT. IV. p. 138.

HAVING treated of the relations of colours, and exemplified them in a distinct essay,* we may speak more briefly on this important branch of our subject, upon which the expression of colours discussed in the preceding section rests, and their right application in practice depends.

BLACK and WHITE are extreme colours, comprehending all other colours synthetically, and affording them all by analysis. The truth of this position is illustrated by our first and second Experiments, Chap. xxiv., wherein the fundamental fact upon which the true natural relations of colours rests is demonstrated by the eduction of an aureola of the three *primary colours* from a black spot upon a white ground, or, *vice versa*, of a white spot upon a black ground, as represented Plate I. fig. 1.

The PRIMARY COLOURS are such as yield others by being compounded, but are not themselves capable of being produced by composition of other colours. They are three only, *yellow, red,* and *blue;*† and are sometimes, by way of distinction, called entire colours.

---
* See "Chromatics, or, an Essay on the Analogy and Harmony of Colours," wherein the relations of each colour to every other colour, and to light and shade, is shown by examples.
† See Note C.

The SECONDARY COLOURS are such only as can be composed of, or resolved into, two primaries, and are also three only; namely, *orange*, composed of red and yellow; *green*, composed of yellow and blue; and *purple*, composed of blue and red.

The TERTIARY COLOURS are such only as can be composed of, or resolved into, two secondary colours, or the three primaries; and these are also three: namely, *citrine*, composed of green and orange, or of a predominant yellow with blue and red; *russet*, composed of orange and purple, or of a predominant red with blue and yellow; and *olive*, composed of purple and green, or of a predominating blue with yellow and red.*

These three genera of colours comprehend in an orderly gradation all the colours which are *positive or definite;* and the three colours of each genus, united or compounded in such subordination that neither predominate to the eye, constitute the negative or NEUTRAL COLOURS, of which *black* and *white* are the opposed extremes, and *greys* are their intermediates. Thus black and white are constituted of, and comprehend latently, the principles of all colours, and accompany them in their depth and brilliancy as shade and light; of which more hereafter.†

Colours thus generally defined are exemplified in the definitive scale of colours, Pl. I. fig. 3, in which their relations to each other, and to light and shade, are distinctly and in an orderly manner discriminated from white to black. It is to be noted, however, that the above denominations of colours do not merely express the individual hues or tints by which they are exemplified in this diagram, but denote classes or genera of colours, each colour comprehending an indefinite series of *shades* between the extremes of light and dark, as each compound colour also does a similar series of *hues* between the extremes of the colours which compose it.

As each class or genus of colours, primary, secondary, and tertiary, has the property of combining in a neutral or achromatic state, when duly subordinated or compounded, it follows that each secondary colour, being compounded of two primaries, is neutralized and contrasted by the remaining primary, alternately; and that each tertiary colour, being a like binary compound of secondaries, is also neutralized or contrasted by the remaining secondary alternately.

These relations of colours will be readily understood by an attentive

---

* See Note D.   † See Exp. I. II. III. Chap. XXVI.

reference to the diagram or SCALE OF CHROMATIC EQUIVALENTS, Pl. I. fig. 2, in which each denomination of colours is opposed to its contrasting denomination reciprocally.

By the term *equivalent* we here denote such a relation of quantities in the combination of antagonist or opposite colours, as produces achromatic or neutral shades, in which the two constituent colours disappear.

Such antagonists or opposites of colours have received different denominations according to circumstances:—thus the spectral antagonist, called an *ocular spectrum*, which arises after long viewing a colour, has been variously denominated *adventitious, accidental,* &c. and is always its true *opposite*, but never its *equivalent*: such colours simply opposed in juxta-position are called *contrasts,* and may be either equivalent or unequal contrasts. All these correspondent colours have also been called *complementary*, although to be properly complementary they ought to be equivalent.

This Scale of Chromatic Equivalents is constituted of six circles, comprehending the primary blue, red, and yellow, and secondary colours, orange, green, and purple, alternately within a larger graduated circle,—the compound denominations appearing within the intersections or crossings of the circles: firstly, the binary, or *secondary compounds,* red-purple, red-orange, yellow-orange, &c. comprising the star formed by the alternate crossings of two circles; and secondly, the ternary, or *tertiary compounds,* russet, &c. comprehended in the smaller central star formed within the crossings of three of the circles alternately.* The graduated scale by which the whole is circumscribed, is divided round the inward edge by numerals diametrically opposed, denoting the proportions in which colours lying on any radius of the circle neutralize and contrast any colour, simple or compound, on the opposite radius; while the mediating colours, which subdue without neutralizing or contrasting, succeed each other side by side all round the scheme: *e. g.,* red subdues and is subdued, or melodized, by the orange and purple contiguous to it, and so on.

The eye is quiet, and the mind soothed and complacent, when colours are opposed to each other in equivalent proportions chromatically, or in such proportions as neutralize their individual activities. This is *perfect harmony,* or union of colours. But the eye and the mind are agreeably moved, also, when the mathematical proportions of opposed or conjoined

---

* See Note E.

colours are such as to produce agreeable combinations to sense; and this is the occasion of the *variety of harmony*, and the powers of composition in colouring. Thus colours in the abstract are a mere variation of relations of the same thing. Black and white are the same colour; and, since colours are mere relations, if there were only one colour in the world there would be no colour at all, however strange, offensive, or paradoxical such assertions may appear.

The neutralizing powers of colours, called *compensating*, have also been improperly denominated *antipathies*, since they are the foundation of all harmony and agreement in colours; too much of any colour in a painting being invariably reconciled to the eye by the due introduction of its opposite or equivalent, either in the way of compounding, by glazing or mingling, or by contrast,—in the first manner with neutralizing and subdued effect, and in the last with heightened effect and brilliancy,—in the one case by overpowering the colour, in the other by overpowering the organ; while in each the equilibrium or due subordination of colours is restored. It is not sufficient, however, that the artist is informed what colours neutralize and contrast, if he remain unacquainted with their various powers in these respects. If he imagine them of equal force, he will be led into errors in practice from which nothing but a fine eye and repeated attempts can relieve him; but if he know beforehand the powers with which colours act on and harmonize each other, the eye and the mind will go in concert with the hand, and save him much disappointment and loss of time, to say nothing of the advantage and gratification of such foreknowledge in realizing their beauties with intention.

We have been enabled to demonstrate the proportional powers of colours numerically, as given in our *Scale of Chromatic Equivalents*, by means of the METROCHROME,[*] whereby it is ascertained that certain proportions of the primary colours, which reduced to their simplest terms are as 3 yellow, 5 red, and 8 blue, of equal intensities, neutralize each other, integrally, as 16; consequently, red 5 is equivalent to green 11, yellow 3 to 13 purple, and blue 8 to 8 orange. The intermediate proportions all round the scale may be obtained by adding any number thereon to that preceding or following it, and the like compounded number diametrically opposite it will be its proportional for the colours on the same diameter. Some of these may

[*] For the principle and mechanism by which we have effected this, see Chap. xxv. Exp. XXVII.

be reduced to simpler terms: thus the equal proportionals 8, to both which the two ends of the needle or index point on the scale, are as unity, 1=1,— the simplest of all ratios, that of equality; and its colours are *orange* and *blue*,—literally the points of extreme *hot* and *cold*, which are, so to call them, the *poles* of harmony in colouring. This result is accidental, but it is a coincidence which evinces the truth of our process, and singularly comports with the rule of harmony in painting, which has been founded on sense or feeling, and requires that equality or balance of warm and cool colouring in a picture upon which tone so essentially depends.*

These are the only two contrasting colours which, like black and white, are equal powers: all other contrasts are perfect only when one of the antagonist colours predominates according to the proportions marked upon the scale. A line diagonally across the needle, or index, indicates the positions of the scale at which colours become most *advancing* and most *retiring;* and a like line perpendicularly across the scale points out all the *middle colours*. These three lines divide the entire scale into equal portions throughout.

Again, by the Scale of Chromatic Equivalents may be determined the proportions in which any *three* colours neutralize and harmonize each other: thus, as 3, 5, and 8, are these proportions of the primaries, yellow, red, and blue,—so 8, 11, and 13, are those of the secondaries, orange, green, and purple. For the more readily finding the proportions of any three harmonizing colours on the scale, it is graduated all round, and divided into three equal parts, which are each subdivided into 32 degrees, numbered accordingly on the outward edge of the scale, and trisecting it all round, so that each colour of the scheme with its two harmonics are indicated by the same number, and the numbers corresponding on the inward edge show their proportions. In like manner may be found the proportions of six or nine harmonizing colours, &c. By causing this external circle of figures to move round the scale, it may be made to indicate the proportions of any number and variety of hues which harmonize; but this is unimportant for practice. This scheme is also a key to the whole science of nature in the painting of

---

* This balance and insensible union of hues and shades in painting, and of tones in music, the Greeks denominated by the same term, TONOS. " Tandem se ars ipsa distinxit, et invenit lumen atque umbras, differentia colorum alterna vice sese excitante: postea deinde adjectus est *splendor*, alius hic quam lumen; quem, quia inter hoc et umbram esset, appellaverunt τόνον:" says PLINY, l. xxxv. c. 5.

flowers, and it coincides therewith that the archetype of all floreal forms is triadic, consisting of the involution of triangles variously irradiated; the numbers of their rays or leaves being invariably 3, 4, or 5, or multiples thereof;—of which only by the by.*

By attention to these relations the student may approximate to a just conception of the powers of colours, and, assisted by a good eye, and a knowledge of his materials, may attain to a perfect application of them; by a like attention to these powers of colours the engraver too will be enabled to estimate those due additions of light or shade which may be necessary to compensate for the absence of colours in his performance; or, in other words, to represent them by their exact equivalents of light and shade.

It has been remarked as a common deficiency of young painters, that their figures, though well drawn, have wanted relief, and that the early works of Vandyke, Titian, and other great masters, have had the same deficiency; for the perfect management of light, shade, and colour, upon which relief depends, is ever of latest attainment, and some there are who never attain it. Yet others, unacquainted with the relations and powers of colours, and even wanting natural feeling therein, colour well by creating for themselves an artificial eye for colouring, and building on the taste and science of other masters, by constantly having a finely-coloured picture as a model while painting. This is nevertheless to be approved where science and nature are wanting, and has by persevering practice established both in the same person; for habit is art and second nature.

We may learn farther from these relations of colours, why dapplings of two or more colours produce effects in painting so much more clear and brilliant, than uniform tints produced by compounding the same colours;—why also, as justly remarked by Sir Joshua Reynolds, colours mixed deteriorate each other, which they do more by imperfectly neutralizing or subduing each other chromatically, than by any chemical action or discordance, though the latter is sometimes also to be taken into the account;—they impress too, on the good colourist, not only the necessity of using his colours pure, but that also of using pure colours; nevertheless, pure colouring and brilliancy differ as much from crudeness and harshness, as tone and harmony do from murkiness and monotony, though both these have been confounded by the injudicious.

* See farther, Exp. XXVIII. Chap. XXVI.

The power of colours in contrasting each other agrees with their correlative powers of light and shade, and is to be distinguished from their power individually on the eye, which is one of light alone: thus, although orange and blue are equal powers as respects each other,—as respects the eye they are totally different and opposed; for orange is a luminous colour, and acts powerfully in irritating, while blue is a shadowy colour, and acts much less powerfully, or contrarily, in soothing that organ,—it is the same in various degrees with other colours; these powers resolve, therefore, ultimately into the same principles of light and shade in a sensible or latent state.

There are yet other modes of contrast or antagonism in colouring, which claim the attention, and engage the skill of the colourist. That of which we have spoken is the contrast of *hues,* upon which depends the brilliancy, force, and harmony of colouring;—there is also the contrast of *shades,* to which belong all the powers of the chiar'-oscuro, which, though it is a part, and the simplest part of colouring only, and ought not to be separated from it, ranks as a distinct and is an important branch of painting, yet is the regimen of opposition in colours coincident with that of light and shade, or black and white; all that can be said of the latter may be said of the former, and he who excels in the one is in a considerable degree qualified to surpass in the other: indeed a just practice of light and shade might carry with it the reputation of good colouring, as it did in Rembrandt, while considerable knowledge of colouring, without the chiar'-oscuro, could not obtain the name of colourists for some eminent masters of the Italian schools. A third mode of contrast in colouring is that of *warmth* and *coolness,* upon which depends the toneing and general effect of a picture; besides which there is the contrast of *colour* and *neutrality,* the chromatic and achromatic, or of hue and shade, by the right management of which local colours acquire value, gradation, keeping, and connexion; whence comes breadth, aerial perspective, and the due distribution of grays and shadows in a picture.

This principle of contrast applies even to *individual colours,* and conduces greatly to good colouring, when it is carried into the variety of hue and tint in the same colour, not only as respects their light and shade, but also in regard to warmth and coolness, and likewise to colour and neutrality. Hence the judicious landscape-painter knows how to avail himself of warmth and coolness in the juxtaposition of his greens, as well as of their lightness and darkness, or their brilliancy or brokenness, in producing the most beautiful and varied effects, which spring in other cases from a like management of blue, white, and other colours. These powers of a colour upon itself are highly important to the painter, and conduce to that gratification from fine colouring, by which a good eye is so mysteriously affected.

If we inspect the works of Nature closely, we shall find that they have no uniform tints, whether it be in the animal, vegetal, or mineral creation;—be it flesh or foliage, the earth or the sky, a flower or a stone,—however uniform its colour may appear at a distance, it will when examined nearly, or even microscopically, be found constituted of a variety of hues and shades compounded with harmony and intelligence.

Transparency and *opacity* constitute another contrast of colouring, the first of which belongs to shade and blackness, the latter to light and whiteness;—even contrast has its contrast, for *gradations, or intermedia*, are antagonists to *contrasts or extremes;* and upon the right management of contrasts and gradation depends the harmony and melody, the breaks and cadences, the tone, effect, and expression of a picture; so that painting is an affair of judicious contrasting so far as it regards colour, if even it be not such altogether.

These contrasts may also be variously or totally conjoined; thus, in contrasting any colour, if we wish it to have light or *brilliancy*, we degrade, or cast its opposite into shade;—if we would have it *warm*, we cool its antagonist;—and if *transparent*, we oppose it by an opaque contrary, and *vice versa*, &c.: indeed, in practice, all these must be in some measure combined.

Such are some of the powers of contrast in colouring alone, and such the diversity of art upon which skill in colouring depends. It must not be forgotten, however, that contrasts or extremes, whether of light and shade, or of colours, become violent and offensive when they are not reconciled by the interposition of their media, or a mean which partakes of both extremes of a contrast; thus blue and orange in contrast become reconciled, softened in effect, and harmonized when a broken colour composed of the two is interposed; the same of other colours, shades, and contrasts.

Another important rule which belongs to the consideration of contrasts is, that *that which holds of the one species, holds also of the others:* hence the maxims of the chiar'-oscuro are applicable to contrasting colours; each have their focus,—should each mutually penetrate and diffuse,—be each repeated subordinately, that is, as principal and secondary, and mutually balance each other, &c. So much, indeed, is the management and mastery of colours dependent on the same principles as light and shade, that it might become a point of good discipline, for the perfect attainment thereof, after acquiring the use of black and white in the chiar'-oscuro, to paint designs in contrast; that is, with two contrasting colours only, in conjunction with black and white; for example, with blue and orange, previously to attempting the whole together. Black may even be dispensed with in these cases, because it may be compounded, since the neutral grey and third colours always arise from the compounding of contrasting colours, so that even flesh may be painted in this way; for example, with red and green alone, as Gainsborough is said to have done at one period of his practice. It is thus that one part of an art becomes a mirror to the rest.

By mixing his *colours with white*, the artist obtains what he has appropriately called his *tints;* by mixing *colours with colours*, he obtains compound colour or *hues;* finally, by mixing *colours or tints with black*, he gets what are properly called *shades;* yet these distinctions are very commonly confounded.

The foregoing classification of colours is an arrangement which exhibits a correct genealogy of their hues and shades in a general view, and enables us to comprehend the simplicity of relation which subsists among an infinity of hues, shades, and tints of colour, while it is calculated to give precision to language respecting colours, the nomenclature of which has ever been exceedingly arbitrary, mutable, and irrelative. The names of colours, consisting of terms imposed without general reference or analogy, according to views and fashions ever varying, are for the most part idiomatical and ambiguous in all languages; yet, boundless as is the variety of hues and compounds, the cultivated eye will readily distinguish the degrees of relation in every possible instance, to the preceding denominations of classes.

There are, however, some anomalous popular names of classes, which, being shades nearly allied to the tertiary colours, have been confounded therewith, and being also of great practical importance merit the consideration of the colourist. These denominations comprehend *all the combinations of the primary, secondary, and tertiary colours, with the neutral, black, or shade*, and therefore may be appropriately called SEMI-NEUTRAL COLOURS; and they comport themselves precisely according to the preceding relations and arrangement throughout.

Of the various combinations of *black*, those in which *yellow, orange, or citrine* predominates, have obtained the terms BROWN, &c. A second class, in which the compounds of black are of a predominant *red, purple, or russet* hue, includes the denomination of MARRONE, CHOCOLATE, &c.; and a third class, in which the combinations of black have a predominating hue of *blue, green, or olive*, comprehends the colours termed GRAY, SLATE, &c. Brown, marrone, and gray may therefore be properly adopted as distinguishing appellations of the three classes of semi-neutral colours. *

These colours are of importance in practice, as following, deepening, or shading colours of the primaries, secondaries, and tertiaries, under which they are to be classed, and not to be confounded therewith as legitimate compounds and relatives. Nevertheless, we know an ingenious and eminent artist, who, confounding *shades* with *hues*, and practice with theory, imagines he can always produce his third colours by the addition of black with unusual simplicity, harmony, and force; and a late professor of painting, upon a like supposition, talked of harmonizing his discordant colours by black. The same principle has probably prevailed upon many palettes: it must indeed be that upon which engravings in black are to be coloured, but would require transcendent skill in the painter to escape murkiness; and it may be presumed to have led many of the old masters into obscurity; so that the horror Rubens expressed at *white in shadows* ought powerfully to prevail against *black in colours*. A greater horror than either is a partiality for a particular colour; but, to put these horrors out of question, the painter ought to dread only the improper use of any colour. Black is, however, to be guarded against in another respect, that as colours or hues in pictures vanish or decay, blackness takes their place; and for this some allowance of freshness and force should be made in painting. Nevertheless, the contrary of this is practised, when an artist, as he is too apt to do, looks at Nature with a prejudiced eye, and sees her objects not in their true colours, but of the hues he has seen in pictures, or of the colours he has been accustomed to paint them. Fearing to look upon her impracticably, or to raise himself up to Nature, he gets over the difficulty by deceiving himself, and pulling her down to his own level.

* See Note D.

Black is the absolute unity of the triad of colours, and hence has in a degree a uniting, monotonizing, or harmonizing power upon them, while it softens their discordances by obscuring them, which is the only rational defence of a practice wherein obscurity, monotony, or shade, is substituted for harmony :—the perfection of colouring is, however, to combine harmony with brilliancy, unity with variety, and freshness with force, without violating the truth of nature. In cases, however, where the artist is constrained to represent his principal objects of given local or offensive colours, as in military dresses, &c., or is otherwise compelled by his subject to paint in a difficult key, this power of light and shade over colour stands him as an important aid. By an illegitimate practice in music, similar to the above, the discords of sound are tolerated by the ear when low in the scale, or thrown, unresolved by their proper consonances, growling into shade, the eye and ear being alike more sensible of high and brilliant relations of colours and sounds than of the low and deep.

If any of the preceding denominations of colours or classes should be objected against, others more significant or analogous may be substituted, if there be such, for we contend not for terms, content in the present case if those we have adopted convey our meaning to the reader.

With regard to the *perspective of colours*, or the manner in which they affect the eye, according to position and distance, it is a branch of aerial perspective, or the perspective of light and shade, and both are governed by similar laws. This perspective of light and colours is distinguished from linear perspective, or the perspective of drawing, as drawing is from colouring; and they have progressed alike in the art. The most antient painters seem to have known little of either; and linear perspective was established as science before the aerial, as drawing and composition preceded colouring.

The perspective of colours depends upon their powers to reflect the elements of light,—powers which are by no means uniform; accordingly, blue is lost in the distance before red, and yellow is seen at a distance at which red would disappear;—yet blue preserves its hue better than red, and red better than yellow, because colours are cooled by distance. In this respect, the compound colours partake of the powers of their components, according to a general rule, by which colours nearly related to black are first lost in the distance, and those nearly related to white disappear last. It is the same with light and shade, the latter of which is totally lost at great distances;

and hence it is that the shadowed side of the moon is not seen. These powers of colours are, however, varied by mist, air, altitude, and mixture, which produce evanescence, and by contrast, which preserves the force of colours by distinguishing them. Colours do not decline in force so much by height as by horizontal distance, because the upper atmosphere is less dense and less clouded with vapour; and hence it is that mountains of great elevation appear much nearer than they really are. From all these circumstances it is evident, that it is not a simple scumbling or uniform degradation of local colours that will effect a true perspective therein; for this will be the aerial of light and shade only; but such a subordination of hues and tints as the various powers of colours require, and is always observable in nature. Few artists have succeeded satisfactorily in this species of effect, which is principally attainable by close application to nature; and we have no finer examples thereof in art than the landscapes of Wilson afford.

If the expression of which we have treated in the preceding chapter may, according to no improper analogy, be called the *poetry of colouring*, the relations which are our present subject have equal claim to be considered as *its music*, for they are that upon which the *harmony of painting* depends, and all ages have consented to class these arts as sisters, and of the same parentage. This affinity of the sister arts, and the unquestionable identity of their archetype, discloses numerous analogies, through which the lights of either art may be reciprocally reflected upon the obscurities of the others, as well in theory as in practice; and it is evident that the hues and combinations of colours are as infinite as those of sounds; and there is hence equal scope for the fine sense and genius of the colourist and musician. Beautiful, fine, distinct, and classed colours are as necessary to the former as sounds of similar qualities are to the latter. The palette is the *instrument* of the painter, as the *viol* is of the musician, and the tone and tuning of the latter is analogous to the colours and setting of the former; each requires such adjustment according to the principles of its respective art. It is difficult indeed to say where this analogy ceases, if it be true, as philosophers have argued and poets sung, that—

> *Nature*, which is the vast creation's soul,
> That steady, curious agent of the whole,—
> The art of heav'n, the order of this frame,—
> Is only *Music* in another name.
>
> CATHARINE PHILLIPS.

In furnishing or *setting the palette* philosophically, harmonically, and upon principle, according to the preceding relations, it is necessary to supply it with pure *blue, red,* and *yellow;*—to oppose to these an *orange* of a hue that will neutralize the blue—*green* of a hue that will neutralize the red—and *purple* of a hue that will neutralize the yellow; and so on to *black* and *white* that also neutralize each other. As in nature the general colour of the sky is blue, and the colour of light is always opposite to that of the sky and shade, the white which is to represent light should be tinged with the orange of the palette sufficiently to neutralize the predominant coldness of black; and pure neutral white may thus be reserved as a local colour, by which term is understood, technically, the natural colour of an object unvaried by distance, reflection, or any other circumstance interfering with distinct vision.

This principle of setting a palette requires that all the colours should be as much as possible in contrast and accordance; and a palette thus tuned or set will afford facilities and conduct to harmonious colouring in ways as various, under the eye of taste and judgment, as the melodies and harmonies which music can elicit, through the genius of the composer, from a fine-toned and well-tuned instrument. It is admitted, nevertheless, that an able hand, guided by fine sense and a judicious mind, may realize fine effects in either art with irregular and inferior means; and granting also that such setting of a palette is theoretical, it is nevertheless not a false refinement, since it is founded in nature, and assimilates with the most approved general practice, and the constant aim of the most eminent colourists; we offer it therefore only as consequent to the relations of which we have been treating.

Such observances may notwithstanding, we repeat, be held unnecessary to the few in whom the perceptions of sense are so powerful, that, with little of the aid of intellect and less of science, they operate and produce with the grace and certainty of nature, as it were by instinct, compared with which the productions of science are mere mimickry. This is the inspiring principle called *genius,* in which we recognise Divinity; to be fraught with which in perfection, is to be above rule,—to mark an æra, and to give canons to art. But though indisposed with some to deny such principle altogether, we admit it to be, in perfection, of the rarest occurrence only, and hold that most excellences of human art have been the joint results of application and rule or science;—or, in other words, of true theory and assiduous practice combined. It has hence been objected that rules are but fetters to genius; and it may be remarked that they enable men of little ability to produce considerable works mechanically; and both these have some truth when rules are used without their reasons;—but reason does not fetter genius; on the contrary, genius itself is latent reason operating by natural rules unconsciously.

CHAP. IV.

ON THE PHYSICAL CAUSES OF COLOURS, &c.

"The mind of Leonardo was, however, too active and capacious to be contented solely with the practical part of his art; nor could it submit to receive as principles, conclusions, though confirmed by experience, without first tracing them to their source, and investigating their causes, and the several circumstances on which they depended."

HAWKINS'S LIFE OF DA VINCI, p. 19.

IN treating of pigments, there is no speculation more natural than the inquiry, *What are the physical causes of colour in general?* We shall not therefore pass it over unnoticed, though it is a question of undoubted difficulty, which has never been satisfactorily answered; being perhaps too abstract and elementary to admit of sensible demonstration, and therefore incapable of popular solution. Hence the theory of colours has varied with the changes of philosophy, and the fashions of the mind.

According to common apprehension, colours are inherent or substantial qualities of bodies, or material things.* The more speculative have ascribed them to sense or vision only, regarding them as intellectual affections through the organs of sight; while a third class of inquirers has assigned them variously a medial station between vision and coloured objects; viz. in light.

Each of these doctrines has been sanctioned by great authorities, and they have given birth to the various hypotheses upon which the theories of colours have been founded and compounded; but, however various and ingenious these theories may have been, none has hitherto fulfilled the strict requisitions of reason or experiment.

It is probable, and perhaps demonstrable, that colours in the abstract belong neither exclusively to the object, the subject, nor the medium, but

---

* It may perhaps be ultimately found, in this as in other cases, that nature does not play the fool with our senses, but that the last accomplishments of science coincide with common apprehension.

refer in equal relation to them all; while they are capable of being regarded in either respect; physically regarded, they are *material objects,*—metaphysically, they are *intellectual impressions,*—and, in an æsthetical sense, they are *sensible representations of material objects to the mind.*

Our present question is however purely physical, or that of the natural philosopher, in which view colours are either INHERENT, as belonging to their passive objects, as in pigments, &c. or TRANSIENT, as belonging to their concurrent medium, as in light, ocular spectra, &c. And in both they are here to be regarded as *material* and of the same nature; but the colours of light being the most pure, simple, and elementary, have afforded the most usual and elegible basis for this inquiry.

Yet sensible LIGHT is not a simple substance, but an effect of the concurrence of two elementary powers; one of which is the active principle of light; the other, passive or re-active, and to be regarded as the principle of shade, or darkness,—the first, coincident if not identical with the *oxygen* of the chemist; the other, with *hydrogen;** and, however exceptionable this may be to those who have been accustomed to regard darkness as a mere privation of light, yet as respects the artist, a principle of darkness, blackness, and shade, is as essential as is a principle of light.

Accordingly the sunbeam, as it arrives to us is a compound of these elements of *light* and *shade*, and it may be analyzed by refraction, and in other ways, into oxydizing or *whitening* rays, and hydrogenizing or *blackening* rays; and at the same time into others that are variously compounded of these, and variously *coloured*.

Light hence appears, as before remarked, to be in the sunbeam the effect of the concurrence or conjugation of two æthereal, electrical, or elementary substances or powers; the one, an agent, of which the sun appears to be the fountain or source; the other, a re-agent, existing in planetary or atmospheric space, analogous to shade;—if so, the sun's light is a species of

---

* We adopt these terms for the two principles of light, not for their fitness, but because they have already been used in a similar elementary sense by the chemists;—otherwise either electrogen and thamogen, or *phosphogen* and *sciogen*, were more analogous terms; and it might be well that natural philosophers should agree upon general appellations for the two opposed or concurrent principles of light,—of electricity,—of galvanism,—of magnetism, and of every chemical or physical elementary science, since there can now be little doubt of their original identity; the experimental demonstration of which, and of our physical rationale of colours, is beside our present purpose, and pregnant with materials for volumes.

oxidation or combustion, a sort of flame attended by sensible or latent heat.

Light has nevertheless been considered, equally by the common and philosophic observers, as single or simple, and not as containing in itself any antagonist principles, but as having merely intension and remission.

Newton was the first who taught to regard the sunbeam as a compound of rays of various powers and colours, but still he regarded it singly, and its heat as accidental. Subsequent investigation has shown that his analysis was chemically defective;—he erred also in regarding light as a compound of heterogeneous rays, of colours and powers essentially different; and in other respects, important to art on account of the great authority of his name.[*]

Scheele and others have demonstrated rays of an invisible kind, accompanying the colours of the sunbeam in the prismatic spectrum, which have been denominated variously *deoxidizing, chemical,* and *phlogistic,* or *hydrogenizing* rays,[†] and prove the existence of a tenebrous or dark principle, by which light is modified. Herschel has also investigated the calorific rays, or heat of the sunbeam; but these are to be regarded, with Newton, rather as accidental, or an effect, than as a constituent or principle of light; thus we have real and demonstrable principles in place of hypothesis, upon which to explain the various phenomena of light and colours.

We may therefore regard *the transient colours of refracted light, and also light itself, as* OXIDES OF HYDROGEN, produced by a species of combustion, attended by heat or caloric, as observed in the sunbeam and prismatic spectrum.

So also are the *inherent colours of solids and liquids* to be regarded upon the same analogy as *oxides of hydrogen,* or, what is the same, as of oxygen united with a phlogistic or inflammable principle. And thus the physical cause of all colours is to be explained upon the same elementary principle or reasoning. All substances too, whether solid, liquid, or elastic, are *attractive or repulsive* of oxygen and hydrogen,—of *one* or *both*,—or they are *neutral;* and all substances are coloured. Hence the affinities of light determine it either to be wholly or partially reflected, transmitted, or refracted, or to enter into chemical combination with material substances.

---

[*] See Exp. I. II. and particularly Exp. XIII. Chap. XXVI.
[†] See Crell's " Journal," vol. III. p. 202. and Scheele's " Essays," p. 206.

From the foregoing constitution and properties of light, and a wide experimental induction, we infer the following propositions:

1. Neutral substances, or such as are in a state of indifference, *neither attractive nor repulsive* of the principles of light, are transmissive, or TRANSPARENT and ACHROMATIC, or colourless.*

2. Substances *entirely repulsive* or reflective of the oxygenous and hydrogenous principles of light, are WHITE and OPAQUE.

3. Substances *entirely attractive*, or absorbent of, or having entire affinity for both principles of light, are BLACK and OPAQUE.

4. Substances having *partial and equal affinities* attractive or repulsive for both principles of light, according to the proportions in which they constitute light, are partially transparent or opaque; i. e. SEMI-PELLUCID and COLOURLESS, or grey.

5. If substances have *unequal affinities* for these oxygenous and hydrogenous principles, they are COLOURED and transparent, or opaque, according to the above conditions.

6. If, in consequence of this unequal affinity, a substance reflect, refract, or transmit light with *one proportion* of the hydrogenous principle in defect, it will be YELLOW—if with *a second proportion*, less deficient, it will be RED—and if with *a third proportion*, but in excess, it will be BLUE; and of proportions intermediate, or compounded of these, will be constituted the secondary and intermediate colours, &c.

The relative proportions in which the primary colours combine in light, &c. in an achromatic or colourless state, as determined by the Metrochrome and denoted by the Scale of Equivalents, is approximately *three* of yellow, *five* of red, and *eight* of blue; † and since the two first belong to one extreme of the prismatic spectrum wherein the hydrogenous principle is in defect, and together amount to *eight*, and the last belonging to the other extreme wherein the hydrogenous principle is in excess, is also *eight*, it appears that the two principles of light are equal and complementary powers.

---

* Hitherto we have had no physical or rational explanation of transparency: every mechanical arrangement of parts is quite insufficient to account for this phenomenon. Transparency and opacity are entirely relative, there being no substance absolutely transparent or opaque. Glass and adamant reflect, and gold transmits, light and colours.

† Exp. XXVII. Chap. XXVI.

Such in briefness we take to be the chemical constitution and physical causes of light and colours, upon which their chromatic relations and effects depend; and our doctrine is illustrated and supported by many facts and experiments. Thus, in the *oxygenation* or oxidisement of metals, which have been not unaptly regarded as compounds of *hydrogen*,* as well as in other inflammables also, the inferior degrees of oxidisement produce *blacks, blues, greens,* &c., but the higher degrees produce *red, yellow, white,* &c.; not uniformly indeed, but generally according to the unknown constitution of the bases of these inflammables themselves. So also in the colours of flame arising from hydrogen and other inflammable substances burning in air or oxygen, we observe at the base of the flame, in which the *hydrogen* abounds, colours tending to *blue;* and toward the apex of the flame, where it is more oxygenated, its colours tend to *yellow;* between which two colours lie tints abounding in *red.* Our principle of the evolution and absorption of the oxygenous element goes also to explain those changes of colour arising and disappearing, which take place by changes of temperature, or simply by heating and cooling, wetting or drying; changes which take place in many pigments, sympathetic inks, &c.

So again in the general and more permanent changes which pigments and colours undergo, oxygen bleaches or fades them, and hydrogen and inflammables deepen or darken them; while light and air, containing both principles, effect both these kinds of change variously, according to affinities already spoken of. The colours of all organic bodies, even to the plumage of birds and insects, depend upon the same principles, and we have found the same chemical effects uniformly in each of these subjects.†

Upon the same principles may be easily explained the production or evolution of the *transient colours* of refracted light, &c. Thus oxygen and hydrogen, having different affinities or activities in the luminous compound, are unequally affected or resisted in passing through transparent bodies, according to their various constitution; and consequently they are unequally refracted,—the oxygenous or more active principle being less so than the

---

* Davy's "Elements."

† Indeed the first principles of light, in an extreme chemical view, seem to be the elements of all material things under different denominations, in remarkable accordance with the first work of creation, as recorded in Genesis c. 1. vss. 2, 3, and 16, in which *darkness* and *light* were as principles before the sun was created.

hydrogenous; and being thus variously dispersed and compounded, they produce colours, as we see, in passing through prisms and lenses,* &c.

Upon this chemistry of light we may easily account for the variety of colours so beautifully displayed in vegetal nature, and principally in flowers, which acquire their colours as they expand, and undergo all the relative changes of hue and tint in their progress and decay, which the immediate combination of these chemical agents may be made to produce in the elaboratory of the chemist.

VISION itself, and its various phenomena, physically explained, are to be regarded as dependent on the same subtile chemistry; for nature is ever simple and uniform. If hence the eye, by the agency of the retina and optic nerve, have *equal affinity for both elements* of light, it will at once discern their intensity or power in light and shade, and also their inequalities in colours; while unequal affinities of the organ may explain the various defects and disorders of vision with regard to colours,† and the use and abuse of coloured spectacles as remedies thereof. The true physical cause of sensation in either of its organs being once established, we shall have an unfailing clue to the cause of all sensation, and to the true physical foundation of their respective sciences.

As the principles of external light exhaust the principles of light in the eye, it is worthy of our attention that the action of direct light, or strongly reflected light, upon the organ, is destructive of vision, temporarily or for ever; and as the angles of reflection and incidence in light are co-equal, it is obvious that we see objects to most advantage for vision and the organ, by avoiding the angle of incidence, or by a position half turned from the light. It is true that in cases of extreme action of light on the eye, instinct and nature lead or compel us to these observances, but we are too inattentive to them in other cases, as in reading, writing, drawing, &c.; and thereby a contrary habit becomes established, till injured vision or blindness ensue.

This theory of vision affords also a solution of the curious phenomena of OCULAR SPECTRA, in which the eye discerns those adventitious or accidental colours, first treated of by Dr. Jurin,‡ and subsequently by Buffon

---

* See Note F.   † See Note G.

‡ In Smith's "Optics." Kircher also has noticed an effect of this kind in his "Ars Magna Lucis et Umbræ," p. 118.

and others, which have no apparent cause out of the organ itself; for the equal affinity of the eye for the two principles of light and colours, is of course destroyed by the action of a colour in which either of them predominates, the predominant principle neutralizing or exhausting its opposite principle in the organ, while its other principle therein continues free, or accumulates during the act of intent vision; and therefore the organ decomposes, by a due election, the light of other objects to which it is impelled to wander, till the balance of principles in the organ itself is restored.* Accordingly the adventitious colour or spectrum occasioned by an object intently viewed is always of its opposite hue, or that compensatory or harmonic colour which restores the equilibrium or neutrality of light and vision; and that the eye does this by *secreting* these principles of light, and retaining them in a latent state, is evinced by the light and dark circles which arise when the side of the closed eye is pressed, by the spark elicited upon the puncturing of the retina in the operation of couching, and also by the extreme sensibility of the eye to light after being long secluded from it. We may thus easily account for that temporary blindness which follows gazing on the sun or a powerful light, by which the principles of vision become exhausted. Thus also that kind of nyctalophia which occurs in tropical climates, which comes on regularly at the close of day, and goes off when it advances, appears to arise from defective secretion or excessive exhaustion, being usually attributed either to disorder of the digestive organs or the power of the sun's light, and is cured or relieved by secluding the eyes from light during the day;—while that kind of nyctalophia, called moon-eyed, which is common to the Bushmen of Southern Africa, who sleep during the day, and are blind when the sun shines, but who, like feline animals, see well in seeming darkness, may be supposed to arise from redundant secretion or defective excretion of the principles of light and vision.

We may in like manner explain also that morbid sensibility of the retina, in which hot colours and strong contrasts become intolerable to the eye, as owing to unequal secretion or an inflamed state of the organ. We perceive herefrom why also the brilliancy of colours declines upon long viewing them, particularly in a strong light. In fine, the physiologist may hence take a hint toward a physical explanation of all sensation in the nervous system, and of the union of all sensible impressions

* See Note H.

in the sensorium or brain, as a link of identity or connexion between the physical and metaphysical world.

It is worthy of remark here, that the phosphori in general emit light of the same colour as that to which they have been exposed.

Newton remarked also, that inflammable or *hydrogenous* substances refract light more powerfully than other substances, and that the diamond does so most of all, whence he framed the admirable conjecture, since proved, that the diamond itself was inflammable. By a like analogy we may infer, that since non-inflammable substances refract light but weakly, and with faint colours, it is probable that the *oxygenous* principle predominates therein, or that they are oxides, which accords with the discoveries of Davy. But much of this is beyond the purposes of painting.

Colour, and what in painting is called transparency, belong principally we perceive to shade; and the judgment of great authorities, by which they have been attached to light as its properties merely, has led to error in an art to which colour is pre-eminently appropriate;—hence the painter has considered colour in his practice as belonging to light only, and hence many have employed a uniform shade tint, regarding shadows only as darkness, blackness, or the mere absence of light, when in truth shadows are infinitely varied by colour, and always so by the colours of the lights which produce them. We must, however, avoid conducting to a vicious extreme, while we incline attention toward the relation of colour to shade, both light and shade being in strictness coessential to colour; but, as transparent, colour inclines to shade, and as opaque it partakes of light, yet the general tendency of colour is to transparency and shade, all colour being a departure from light. It hence becomes a maxim, which he who aspires to good colouring must never lose sight of, *that the colour of shadow is always transparent, and that of extreme light objects only opaque.*

For this we have also the high authority of Rubens, who in the following extract from his Lessons says, " Begin by painting in your shadows lightly, taking particular care that no white is suffered to glide into them; it is the poison of a picture, except in the lights: if once your shadows are corrupted by the introduction of this baneful colour, your tones will be no longer warm and transparent, but heavy and leady. It is not the same in the lights; they may be loaded with [opaque] colour as much as you may think proper, provided the tones are kept pure: you are sure to succeed in placing each tint in its place, and afterwards by a light blending with the brush or pencil, melting them into each other without tormenting them, and on this preparation may be given those decided touches which are always the distinguishing marks of the great master."

It is to be noted also, that the colour of shadow is always complementary to that of its light, modified by the local colours upon which it falls; and this accords equally with correct observation and the foregoing principles, although it is often at variance with the practice of artists.

Of the mechanical, dynamical, or optical relations of light and shade, so far as regards painting and colours, we need only briefly remark, that the motion or action of light is either *direct, reflected,* or *inflected;*—that the DIRECT LIGHTS of the sun and moon are always in straight lines parallel to each other;—that artificial lights diverge from themselves as centres in radii, and all light partakes of the colour of the medium through which it passes;—that of REFLECTED LIGHT, the angles of reflection are always equal to the angles of incidence, and partake of the colours of the reflected surfaces;—and, respecting REFRACTED LIGHTS, that in passing through transparent media, or by opaque objects, light, whether direct or reflected, is always inflected with a developement also of much or little colour;—and that the shadows of light in each case is always the chromatic equivalent of such light.

In passing an opaque object, light is always bent or inflected toward or into its shadow, and the shadow bends into the light; consequently there is a penumbra surrounding every shade, forming a softening medium between it and the light, and aiding reflection in enlightening every shade; every light has hence its shade, and every shadow its light. It is in the management of these properties of light that the skill of the artist is no less requisite and conspicuous than in the management of colours, with which they are intimately connected.

To conclude,—we know not whether the preceding attempt to explain the causes and effects of vision, light, and colours, physically or chemically, may prove satisfactory to other minds; but of this we feel assured, that the first elements of things are powers and not particles; that the modern corpuscular and undulatory doctrines, with all the mathematical and mechanical explanations hitherto employed, are entirely incompetent to the solution of these phenomena; and that all the hypotheses built upon them, like those which they superseded,* must ultimately fail at the foundation, not even answering the inquiry of the poet :—

> Why does one climate and one soil endue
> The *blushing poppy* with an *orange hue*,
> Yet leave the *lily pale*, and tinge the *violet blue?*
>
> PRIOR.

* Of the best of these was that often quoted of Empedocles, the Pythagorean, from which school has emanated some of the most refined and important of ancient and modern systems;— but with this hypothesis, which accounted for vision by the emission of light from the eye, the powerful mind of Socrates declared itself not thoroughly satisfied.—See Sydenham's Plato; Meno, p. 75.

CHAP. V.

ON THE DURABILITY AND FUGACITY OF COLOURS.

> Parthenius thinks in Reynolds' steps he treads,
> And ev'ry day a different palette spreads;
> Now bright in vegetable bloom he glows,
> His *white*—the *lily*, and his *red*—the *rose*;
> But soon aghast, amid his transient hues,
> The ghost of his departed picture views:
> Now burning minerals, fossils, bricks, and bones,
> He seeks more durable in dusky tones,
> And triumphs in such permanence of dye,
> That all seems fix'd, which time would wish to fly.
>   SHEE.

IN the preceding discussion colours are distinguished into *inherent* and *transient*, the latter of which, as their name implies, are essentially fugaceous; our present argument is, therefore, limited to the permanence and mutability of the inherent colours of pigments, as those which are principally important to the artist.

All durability of colour is relative, because all material substances are changeable and in perpetual action and reaction; there is therefore no pigment so permanent as that nothing will change its colour, nor any colour so fugitive as not to last under some favouring circumstances; while *time* of short or long continuance has generally the immediate effect thereon of *fire* more or less intense, according to the laws of combustion and chemical agency. It is indeed some sort of criterion of the durability and changes of colour in pigments, that time and fire produce similar effects thereon:— thus if fire deepen any colour, so will time;—if it cool or warm it, so will time;—if it vary it to other hues, so will time;—and if it consume or destroy a colour altogether, so also will time ultimately; but the power of time varies extremely with regard to the period in which it produces those effects which are instantly accomplished by fire :—fire is also a violent test, and subject to many exceptions.

That there is no absolute but only relative durability of colour may be proved from the most celebrated pigments;—thus the colour of ultramarine, which, under the ordinary circumstances of a picture, will endure a hundred centuries, and pass through naked fire uninjured, is presently destroyed by the juice of a lemon or other acid. So again the carmine of cochineal, which is very fugitive and changeable, will, when secluded from light, air, and oxygen, continue half a century or more; while the fire or time which deepens the first colour will dissipate the latter altogether. Again, there have been works of art in which the white of lead has retained its freshness for ages in a pure atmosphere, and yet it has then been changed to blackness after a few days' or even hours' exposure to a foul air. These and other affections of colours will be instanced throughout when we come to the consideration of individual pigments; not for the purpose of destroying the artist's confidence in his materials, but as a caution and guide to the availing himself of their powers properly.

It is therefore the lasting under the ordinary conditions of painting, and the common circumstances to which works of art are exposed, which entitle a colour or pigment to the character of permanence; and it is the not-so-enduring which subjects it properly to the opposite character of fugacity; while it may obtain a false repute for either, by accidental preservation or destruction under unusually favourable or fatal circumstances, all of which has been frequently witnessed.

It has been supposed by some that colours vitrified by intense heat are durable when levigated for painting in oil or water. Had this been true, the artist need not have looked farther for the furnishing of his palette than to a supply of well-burnt and levigated enamel colours;—but though these colours for the most part stand well when fluxed on glass, or in the glazing of enamel, porcelain, and pottery, they are almost without exception subject to the most serious changes when ground to the degree of fineness necessary to render them applicable to oil or water painting, and become liable to all the chemical changes and affinities of the substances which compose them. These remarks apply also to those who ascribe permanence to native pigments only, such as the coloured earths and metallic ores.

Others, with some reason, have imagined that when pigments are locked up in varnishes and oils, they are safe from all possibility of change; and there would be much more truth in this position if we had an impenetrable varnish,—and even then it would not hold with respect to the action of light, however well it might exclude the influences of air and moisture; but in truth varnishes and oils themselves yield to changes of temperature, to the action of a humid atmosphere, and to other chemical influences: their protection of colour from change is therefore far from perfect; and the above opinion of them is only in some degree true, but ought not to render the artist inattentive to the durability of his colours in themselves. Reynolds unfortunately entertained this opinion of the preserving power of varnishes; and although the practice of his own palette was exceedingly empirical, he was an utter condemner of such practice in others.*

On the other hand, want of attention to the unceasing mutability of all chemical substances, and their reciprocal actions, has occasioned those changes of colour to be ascribed to fugitiveness of the pigment, which belong to the affinities of other substances with which they have been improperly mixed and applied. It is thus that the best pigments have sometimes suffered in reputation under the injudicious processes of the painter, and that these effects and results have not been uniform in consequence of a desultory practice. If a pigment be not extremely permanent, diluting it will render it in some measure more weak and fugitive; and this occurs in several ways,—by a too free use of the vehicle, by complex mixture in the formation of tints, and, by distribution, in glazing colours upon the lights downward, or scumbling colours upon the shades upward, &c.

The foregoing circumstances, added to the variableness of pigments by nature, preparation, and sophistication, have often rendered their effects equivocal, and their powers questionable; all which considerations enforce the expediency of using colours as pure and free from unnecessary mixture as possible; for simplicity of composition and management is equally a maxim of good mechanism, good chemistry, and good colouring. Accordingly, in the latter respect, Sir Joshua Reynolds gives it as a maxim, that the less colours are mixed, the brighter they appear; the causes of which we have mentioned already. His words are: "Two colours mixed together will not preserve the brightness of either of them single, nor will three be as bright as two: of this observation, simple as it is, an artist who wishes to

---

* Northcote's "Memoirs of Sir J. R."—Supplement, p. LXXX.

colour bright will know the value."—Note xxxvii. to Dufresnoy's Art of Painting.

There prevail, notwithstanding, two principles of practice on the palette, opposed to each other—the one, simple; the other, multiple. That of simplicity consists in employing *as few pigments, &c. as possible;* according to the extreme of which principle the three primary colours are sufficient for every purpose of the art. This is the principle of composition in colouring, the opposite of which may be called the principle of aggregation, and is in its extreme that of having *as many pigments, if possible, as there are hues and shades of colour.*

On the first plan *every tint requires to be compounded;* on the latter, *one pigment supplies the place of several,* which would be requisite in the first case to compose a tint;—and as the more pigments and colours are compounded, the more they are deteriorated or defiled in colour, attenuated, and chemically set at variance, while original pigments are in general purer in colour as well as more dense and durable than compound tints, there appear to be sufficient reasons for both these modes of practice; whence it may fairly be inferred, that a practice composed of both will be best, and that the artist who aims at just and permanent effects should neither compound his pigments to the dilution and injury of their colours, when he can obtain pure intermediate tints in single, permanent, original pigments, nor yet multiply his pigments unnecessarily with such as are of hues and tints he can *safely* compose extemporaneously of original colours upon his palette. This will require experience; and to facilitate the acquisition of such experience is one of the objects of this work.

Examples are to be found of each of these modes in the practice of the most eminent artists; and if the records left us of their palettes prove the fact, the mode of Rubens, Teniers, Hogarth, and Wilson, was more or less that of simplicity. With respect to Sir Joshua Reynolds, we have been assured by his favourite pupil, the venerable Northcote, who was every way interested in remarking and remembering his methods, that however he might have been betrayed by his materials, his practice in using them was regulated by that breadth, simplicity, and generality which marked his great mind, and that hence he placed no expletive tints upon his palette, nor did he torture his colours with the palette-knife or pencil: by which judicious practice, his pictures, notwithstanding partial failures, have triumphed over the imperfection of his materials, and such of his works as have been preserved will remain to posterity with a permanence of hue not inferior to those of any of his great predecessors; while there is a grace and refinement in all his productions, that insure lasting esteem to those even of which the colouring may have partially flown, owing to the employment of the carmine of cochineal and orpiment with blue-black in the formation of his tints; which three, with black and white, constituted the ordinary setting of his palette, till he was forced very unwillingly to give up the two first for vermilion and Naples yellow, which he afterwards continued to employ as long as he painted. Rubens's advice to his pupils, preserved by the Chevalier Mechel, is that of the utmost simplicity; thus, in the painting of flesh, he says: "Paint your high lights *white;* place next to it *yellow*, then *red*, using *dark red* as it passes into shadow; then with a brush filled with *cool gray*, pass gently over the whole until they are tempered and sweetened to the tone you wish."

Vandyke's practice was similar to his master's; and such also was that of Correggio, who painted his flesh with the three primary colours, loaded or embossed his lights, and moderately softened them into his mezzotints, carefully preserving his shadows uncontaminated with white. There is indeed ever something in simplicity which associates it with grace, truth, beauty, and excellence:

> O sister meek of Truth,
> To my admiring youth
> Thy sober aid and native charms infuse!
> The flow'rs that sweetest breathe,
> Though beauty cull'd the wreath,
> Still ask thy hand to range their order'd hues.
> COLLINS, ODE TO SIMPLICITY.

The practice of Sir Thomas Lawrence and the late Mr. Owen afford examples of the aggregate mode, which is well suited also to the painters of flowers and subjects of natural history; and many of the ablest living artists practice in the middle mode, which is certainly less dangerous in respect to permanence.

The first of these three plans, it is true, is the most scientific, since it depends upon the mind, and a thorough knowledge of the relations and effects of colours; while the second depends wholly upon the eye, and is simply the method of sense. This distinction applies also to the two methods which have prevailed with different artists, by the most ordinary of which the painter blends his colours and forms his tints upon his palette, as was probably done by Titian and the Venetian school; and the other by which he makes as it were a palette of his picture, applying his colours unbroken, and producing his combinations therewith upon his canvas, as appears to have been the method of Rubens and Reynolds. To the first, a good eye is principally necessary; while the latter depends upon a better knowledge of colours,—it is also more favourable to the brilliancy, purity, and durability of colouring, according to the foregoing maxims. Here also the best practice is a compound of the two extreme methods, and in some measure essential to good colouring, and appears to have been acted upon in the landscapes of Gainsborough and Wilson.

The practice of producing tints and hues by *grinding* pigments together, instead of *blending* them on the palette, has fallen into disuse, whether advantageously or otherwise may be questioned; but to this disuse may be attributed some peculiarity of the tints and textures of the pictures of the Flemish school, they being perhaps results of intimate combination from grinding, and consequently of a more powerful chemical action among the ingredients compounded. It conduces also undoubtedly to that union upon which tone and mellowness depend, when the same pigments which lie near together in a picture are employed to form intermediate hues and tints; but this practice conducts to foulness when the colours of such pigments are not pure and true, and they do not assimilate well in mixture chemically.

We could not well avoid this digression on the modes of practice, upon which durability so much depends: for so concurrent is permanence with reputation, and so important is it, that had the sculpture of the Greeks been no more durable than their painting, literature could not have preserved the fame of their artists; and the reputation of their painters in our time is principally conceded to the transcendent excellence of their sculpture which still remains. Hence the modern artist, who is inattentive to the durability of his materials, may content himself with sharing some portion only of the reputation of the engraver, whose art will become to modern times in this respect, what sculpture in relief has been to the antient; for as to colouring in particular, which has been called mechanical and subordinate, it is the only department of painting which cannot be copied and transferred mechanically by a copper-plate and a press, but requires a cultivated taste and judgment, a fine eye, and an able hand, united *immediately* in the work.

Abstractedly considered, it is probable that the durability of colour in substances is uniformly dependent upon the state in which they exist chemically or by constitution, with respect to what we have before regarded as the two concurrent and essential principles of light and colours, which we have distinguished by the terms oxygen and hydrogen; and this may account for the apparent capriciousness by which a pigment is found sometimes durable and sometimes not so, particularly the lakes, carmines, and most vegetal colours, the fugitiveness of which depends as much upon the state of their bases as upon the natural infirmity of their colouring matter. If the pigment or its base be in a state which the chemists have termed a *protoxide*, it will, by a gradual acquisition of oxygen from light, air, or moisture, change or fade, till being saturated, or becoming a *peroxide*, it is no farther subject to change by the election of oxygen.

On the other hand, pigments and bases assume similar states with respect to the hydrogenous principle of light, &c. in which they may be termed *prothydrids* and *perhydrids*, in which states they are subject to changes opposite but analogous to the preceding, and to this latter influence metallic colours and their bases are principally subject: upon the whole, however, oxygen being the more active of the two principles, colours are in a greater measure subject to its influence. Such is the subtile chemistry, in the simplest view we are able to take of it, upon which the changes of colours in pigments depend.

It would not be difficult to explain upon the same agency, why pigments are more subject to oxidation and fading in a water vehicle, and to hydrogenation and darkening in one of oil.

With respect also to permanence, it is worthy of remark, that *recent pigments*, both natural and artificial, like recent pictures, wines, &c. undergo amelioration by the influence of time, temperature, atmosphere, &c. which it is better they should suffer in the state of pigments than upon the canvas. Their drying in oil is in general also improved by age, and they approach nearer altogether to the condition of native pigments.

These effects of time, &c. are additional reasons for the esteem in which the colours as well as the works of celebrated deceased artists are commonly held; and this accords with the common remark, that time effects a mellow and harmonious change upon pictures; but sometimes it produces changes altogether unfavourable. To insure the former, and prevent these latter changes, the attention of the artist, in the course of his colouring, should be directed to the employment of such colours and pigments as are prone to adapt themselves in changing to the intended key of his colouring and the right effect of his picture. For example, if he design a cool effect, ultramarine has a tendency through time to predominate, and to aid the natural key of blue : he will therefore compromise the permanence of his effect, if in such case he employ a declining or changeable blue, or if he introduce such reds and yellows as have a tendency to warmth or foxiness, by which the colouring of many pictures has been destroyed. In a glowing or warm key the case is in some measure reversed,—not wholly so, for it is observable that those pictures have best preserved their colouring and harmony in which the blue has been most lasting, by its counteracting the change of colour in the vehicle, and that suffusion of dusky yellow which time usually bestows upon pictures even of the best complexion.

Newly-discovered pigments, however flattering in appearance or in working, are to be employed with caution, or even suspicion, till experience has obtained them the stamp of excellence. Good pigments have ever been prized with so true an estimation of their value by all people, whether barbarian or refined, that it may be doubted if a really excellent one has ever been lost to the world; and to produce such after the ages of research which have passed, and for which all who have had eyes have been in a measure qualified, is, we may be assured, no ordinary result either of accident or design. Accordingly, most of the resplendent pigments, fruits of the fecundity of modern chemistry, have been found deficient of the intrinsic and sterling excellences which have given value and reputation to some of the antient and approved. Thus the splendid *yellow chromates of lead*, which withstand the action of the sunbeam, become by time, foul air, and the influence of other pigments, inferior even to the ochres. So again *Indian yellow*, which also powerfully resists the sun, is soon destroyed in oil, and changed by time, &c. Again, the vivid and dazzling *reds of iodine* are chameleon colours, subject to the most sudden and opposite changes, and yield the palm of excellence to the fine solid colours of vermilion, whose name they usurp and falsify. So again the brilliant *blues of cobalt*, beautiful and abundant as they are, which resist the sunbeam powerfully and have no present imperfection, are always tending to greenness and obscurity, and must yield their pretensions on the palette to the unrivalled excellences of ultramarine.

These, among the chief productions of modern chemistry, are valuable for the ordinary and temporary purposes of painting; but they captivate the eye by a meretricious beauty which misleads the judgment, and are to be introduced with great caution in the more elevated practice of the art.

As to the individual permanency or fugitiveness of pigments, we have noted them under their respective heads as they occur in the following chapters.

CHAP. VI.

ON THE GENERAL QUALITIES OF PIGMENTS.

"Je voi bien," Damon dit, "que vous voulez que le peintre ne laisse rien échapper de tout ce qui est de plus avantageux dans son art."—DU PILE, DIAL. p. 9.

HITHERTO we have treated of colour only, which is the universal quality of pigments, and of its relations, physical causes, and changes; there remain therefore for discussion the more material properties, upon which depend the various uses, excellences, and defects of pigments.

The general attributes of a perfect pigment are beauty of colour, comprehending pureness and richness, brilliancy and intensity, delicacy and depth,—truth of hue,—transparency or opacity, well-working, crispness, setting-up or keeping its place, and desiccation or drying well; to all which must be superadded *durability* when used, a quality to which the health and vitality of a picture belong, and is so essential, that all the others put together without it are of no esteem with the artist who merits reputation: we have therefore given it a previous distinct consideration.

No pigment possesses all these qualifications in perfection, for some are naturally at variance or opposed; nor is there any pigment that cannot boast excellence in one or more of them. BEAUTY, delicacy, purity, and brilliancy, are commonly allied in the same pigment, as are also depth, richness, and intensity in the beauty of others; and some pigments possess all these in considerable degree; yet delicacy and depth in the beauty of colours are at variance in the production of all pigments, so that perfect success in producing the one is attended with some degree of failure in the other, and when they are united it is with some sacrifice of both;—they are the male and female in beauty of colour; the principle is universal, and the

Hercules, Venus, and Apollo, are illustrations of it in sculpture. Hence the judicious artist purveys for his palette at least two pigments of each colour, one eminent for *delicate beauty*, the other for *depth*. Of the importance of beauty in colours and pigments there can be no dispute, since it is equally a maxim in colouring as of sounds in music, that if individual colours or sounds be disagreeable to the eye or ear, no combination of them can be pleasing either in melody or harmony—succession or conjunction.

TRUTH OF HUE is a relative quality in all colours, except the extreme primaries, in the relations of which blue, being of nearest affinity to black or shade, has properly but one other relation, in which it inclines to red, and becomes a purple blue; it is therefore a faulty or false hue when, inclining to yellow, it becomes of a green hue: but red, which is of equal affinity to light and shade, has two relations, by one of which it inclines to blue, and becomes a purple-red or crimson; and by the other it inclines to yellow, and becomes an orange-red or scarlet, neither of which are individually false or discordant; yet yellow, which is of nearest affinity to white or light, has strictly but one true relation by which it inclines to red, and becomes a warm yellow, for by uniting with blue it becomes a defective green-yellow. Thus greenness is inimical to truth of hue in these primaries, agreeably to the law or regulation by which green is as naturally adapted to contrast as it is inept to compound with colours in general. The other secondary and tertiary colours, having all duplex relations, may incline without default to either of their relatives.

TRANSPARENCY is an essential property of all glazing colours, and adds greatly to the value of dark or shading colours; indeed it is the prime quality upon which depth and darkness depend, as whiteness and light do upon opacity or reflecting power. Opacity is therefore the antagonist of transparency, and qualifies pigments to cover in dead-colouring or solid painting, and to combine with transparent pigments in forming tints; and hence also semitransparent pigments are qualified in a mean degree both for dead colouring and finishing. As excellences therefore, transparency and opacity are relative only—the first being indispensable to shade in all its gradations, as the latter is to light. The natural and artificial powers, or depth and brilliancy, of every colour lie within the extremes of black and white; it follows therefore that the most powerful effects of transparent colours are to be produced by glazing them over black and white: as however few transparent pigments have sufficient body or tingeing power for this, it is necessary rather to glaze them over deep reflective or opaque colours of their own hues.

WORKING WELL depends principally upon fineness of texture and the quality called *body* in colours; yet every pigment has its peculiarities in respect to working both in water and oil, and these must become matters of every artist's special experience; and some of the best pigments are most difficult of management, while some ineligible pigments are rich in body and free in working;—yet accidental circumstances may influence all pigments in these respects, according to the artist's particular mode of operation and his vehicle, upon the affinities of pigments with which depend also their general faculties of working, such as keeping their place, crispness or setting-up, and drying well; but these latter and other qualities and accidents of pigments have little of a general nature, and will be particularly considered in treating of the individual characters of pigments: it may however be remarked, that crispness, setting-up, and keeping their place and form in which they are applied, are contrary to the nature of many pigments, and depend in painting with them upon a gelatinous texture of their vehicle;—thus mastic and other resinous varnishes give this texture to oils which have been rendered drying by the acetate or sugar of lead;—simple water also, albumen, and animal jelly made of glue or isinglass, give the same property to oils and colours; bees'-wax has the same effect in pure oils. White lac varnish, and other spirit varnishes, rubbed into the colours on the palette, enable them also to keep their place very effectually in most instances. This is important also, because glazing cannot be performed unless it be with a vehicle which keeps its place, or with colours which give this property to the vehicle, as some lakes and transparent colours do.

FINENESS OF TEXTURE is gotten by *grinding* and *levigating* extremely, but is only perfectly obtained by *solution*—and this few pigments admit of;—it merits attention, however, that colours ground in water in the state of a thick paste, and others, such as gamboge, in strong solution in water and liquid rubiate, &c. are miscible in oil, and dry therein firmly; and in case of utility or necessity, any water-colour in cake, being rubbed off thick in water, may then be diffused in oil, the gum of the cake acting as a chemical medium of union to the water and oil without injury. And pigments, which cannot otherwise be employed in oil or varnish, may be thus forced into the service, and add to the resources of the painter in oil.

In such case, however, the steel palette-knife should be employed with caution.

With respect to DESICCATION OR DRYING, the well-known additions of the acetate or *sugar of lead*,* *litharge*, and *sulphate of zinc*, called also white copperas and white vitriol, either mechanically ground or in solution, for light colours; and japanner's gold size, or oils boiled upon litharge, for lakes, or in some cases verdigris and manganese for dark colours, may be resorted to when the colours or vehicles are not sufficiently good dryers alone; and it requires attention, that an excess of dryer renders oil saponaceous, is inimical to drying, and injurious to the permanent texture of the work. Some colours, however, dry badly from not being sufficiently edulcorated or washed, and many are improved in drying by passing through the fire, or by age. Sulphate of zinc, as a dryer, is less powerful than acetate of lead, but is preferable in use with some colours, upon which it acts less injuriously; but it is supposed, erroneously, to set the colours running; which is not positively the case, though it will not retain those disposed to move, because it wants the property the acetate of lead possesses, of gelatinizing the mixture of oil and varnish. These two dryers should not be employed together, as frequently directed, since they counteract and decompose each other by double election,—forming two new substances, the acetate of zinc, which is an ill dryer, and the sulphate of lead, which is insoluble and opaque. The inexperienced ought here to be guarded also from the highly improper practice of some artists, who strew their pictures while wet with the acetate of lead, or use this substance otherwise in its crystalline or granular form, without grinding or solution, which, though it may promote present drying, will ultimately effloresce on the surface of the work, and throw off the colour in sandy spots.

It is not always that ill drying is attributable to the pigments or vehicle,—the states of the weather and atmosphere have great influence therein. The oxygenating power of the direct rays of the sun renders them peculiarly active in drying oils and colours, and was probably resorted to before dryers were added to oils, particularly in the warmer climate of Italy, in which the very atmosphere is imbued with the matter of light to which the drying property of its climate may be attributed. The ground may also advance or retard drying, because some pigments, united either by mixing or glazing,

---

* This is the *Saccharum Saturni* of the old chemists, and the *Saturnus Glorificatus* of the alchymist, celebrated for its uses in forming pastes for artificial gems, for drying oils, &c.

are either promoted or obstructed in drying by their conjunction. The best practice in this respect is to sponge the picture previously to painting thereon with soft water, and in damp weather with weak aqueous solution of the acetate or sugar of lead. The various affinities of pigments occasion each to have its more or less appropriate dryer; and it would be a matter of useful experience if the habits of every pigment in this respect were ascertained;—siccatives of less power generally than the above, such as the acetate of copper and the oxides of manganese, to which umber owes its drying quality, and others might come into use in particular cases. Many other accidental circumstances may also affect drying; and among these none is more to be guarded against by the artist than the presence of soap or alkali, too often left in the washing of his brushes, which, besides other ill effects, decompose and are decomposed by acetate of lead and other dryers, and retard drying, in streaks and patches on the painting; in all which cases however the odium of ill drying falls upon some unlucky pigment. To free brushes from this disadvantage, they should be cleansed with the oils of linseed and turpentine.

To all other good qualities of pigments it would be well if we could in all cases add that of being INNOXIOUS;—as this however cannot always be, and good pigments are by no means to be sacrificed to the want of this property, while no pigment that is not imbibed by the stomach will in the slightest degree injure the health of the artist; common cleanliness, and avoiding the habit of putting the pencil unnecessarily to the mouth, so common in water-painting, are sufficient guards against any possibly pernicious effects from the use of any pigment.

CHAP. VII.

ON COLOURS AND PIGMENTS INDIVIDUALLY.

"Parmi les couleurs artificielles le peintre doit connoître celle qui ont amitié ensemble, (pour ainsi dire,) et celle qui ont antipathie; il en doit sçavoir les valeurs séparément, et par comparison des unes aux autres."—Du Pile, Dialogue, p. 6.

Having defined and exemplified colours generally, and discussed briefly their relations, causes, and general attributes, we proceed to the more particular and practical part of our work—*the powers and properties of colours and pigments individually;* a subject so pregnant with materials, and of such unlimited connexions, that volumes might easily be inflated to little purpose with vague instructions to prepare bad colours, while good ones may be obtained at less expense—with the history of antient and modern colours, and the biography, if we may use the term, of individual pigments; while our design is merely to sketch simply and briefly their characters and uses, so as to bring the student to a knowledge of his materials by the shortest course: and this we purpose to do in the order suggested by their relations and the foregoing distribution, under their distinct heads.

In so doing we have introduced more illustrations of the *poetic uses of colours* than might appear necessary; because, in the absence of examples from paintings, or those of nature, they may serve to lead the mind into acquaintance with the expression and powers of colours in the abstract, and to fix them as impressions habitually on the mind of the artist by whose taste and feeling they are to be applied; at the same time we have rendered our quotations as brief as the sense would admit, which in many of the instances is of wider reference than could have been exhibited without swelling our illustration beyond reasonable bounds; yet we would willingly call the attention of the student in the widest reference possible to the poet's art as a powerful auxiliary: antients and moderns have used it as such, and Homer may be considered not the patriarch of poets only, but of painters also. Plutarch remarks, "that *poetry is an imitative art that hath in it much of the nature of painting;*" and he observes, "that it is a common saying that *poetry is vocal painting,* and *painting, silent poetry.*"* The same may be asserted of colouring in particular.

With regard to the *beauty of colours individually,* it is a general law of their relations, confirmed by nature and the impressions of sense, that *those colours which lie nearest in nature to light have their greatest beauty in their lightest tints; and that those which lie similarly toward shade are most beautiful in their greatest depth or fulness,*—a law which of course applies to *black* and *white* particularly. Thus the most beautiful *yellow,* like *white,* is that which is lightest and most vivid; *blue* is most beautiful when deep and rich, while *red* is of greatest beauty when of intermediate depth or somewhat inclined to light,—and their compounds partake of these relations: we speak here only of the *individual beauty* of colours, and not of that *relative beauty,* by which every tint, hue, and shade of colour becomes pleasing or otherwise according to place and reference, for this belongs to the general nature and harmony of colours.

There is however a vicious predilection of some artists in favour of a particular colour, from which some of the best colourists have not been totally free, which arises nevertheless from organic defect or mental association; but these minions of prejudice are greatly to be guarded against by the colourist, who is every way surrounded by dangers: there is danger on the one hand lest he fall into whiteness or chalkiness; on the other, into blackness or gloom: in front, he may run into fire and foxiness, or he may slide backward into cold and leaden dulness: all these are extremes he must avoid. There are also other important prejudices to which the eye is liable, in regard to colours individually, which demand also his particular attention, because they arise from the false affections of the organ itself, to which the best eye is most subject; these are occasioned by the various specific powers of single colours acting on the eye according to their masses,—the activity of light, or the length of time they are viewed; whereby vision becomes over-stimulated, unequally exhausted, and endued, even before it is fatigued, with a spectrum which clouds the colour itself,

---

* Works, vol. III. p. 46.

and gives a false brilliancy, by contrast, to surrounding hues, so as totally or partially to throw the eye off its balance and to mislead the judgment. This may be effected by a powerful colour on the palette, a mass of drapery, the colour of a wall,* or other accidental cause; and the remedy against it is to refresh the eye with a new object, of nature, if possible, or to give it rest. The powers of colours in these respects will be hereafter adverted to under their distinct heads. As to the powers of pigments individually, and their reciprocal action and influence chemically, these will be denoted separately of each colour or pigment, and such colours as injure each other pointed out, leaving it to be understood that in instances not noticed colours may be mixed and employed with impunity.

* See Note I.

CHAP. VIII.

ON THE NEUTRAL, WHITE.

> I take thy hand;—this hand
> As soft as dove's down, and as *white* as it;
> Or *Ethiopian's tooth*, or the *fann'd snow* that's bolted
> By the northern blast twice o'er.
>
> SHAKSPEARE.

WHITE, in a perfect state, should be neutral in hue, with regard to colour, and absolutely opaque; that being the best which reflects light most brilliantly. This is the property in white called *body*, which term in other pigments, more especially in those which are transparent, means *tingeing power*. White, besides its uses as a colour, is the instrument of light in painting, and compounds with all colours, when pure, without changing their class; yet it dilutes and cools all colours except blue, which is specifically cold; and, though it does not change nor defile any colour, it is defiled and changed by all colours. This pureness of white, if it be not in some degree broken or tinged, will cast down or degrade every other colour in a picture, while itself becomes harsh and crude. Hence the lowness of tone which has been thought necessary in painting, but is so only because our other colours do not approach to the purity of white. Had we all necessary colours thus relatively pure as white, colouring in painting might be carried up to the full brilliancy of nature.

The term colour is equivocal when attributed to the neutrals, yet the artist is bound to consider them as colours; and, in philosophic strictness, they are such in extreme composition and latently, for a thing cannot but be that of which it is composed, and neutrals are composed of all colours.

Locally, white is the most advancing of all colours in a picture, and produces the effect of throwing other colours back in different degrees, according to their specific retiring or advancing powers; which powers are not however absolute properties of colours, but dependant upon the relations of light and shade, which are variously appropriate in all colours: hence it is that a white object, properly adapted, appears to detach, distribute, put in keeping, and give relief, decision, distinctness, and distance to every thing around it. White itself is advanced or brought forward, unless indeed white surround a dark object, in which case they retire together. In mixture white communicates these properties to its tints, and harmonizes in conjunction or opposition with all colours, but lies nearest in series to yellow, and remotest from blue, of which, next to black, it is the most perfect contrast. It is correlative with black, which is the opposite extreme of neutrality. We have said that black and white are the same colour; and the truth of this appears practically in painting a white object upon a light ground, which is done with black pigment; and also in painting a black object upon a dark ground, which is done with white pigment: in the latter case, by supplying the lights of the object; and in the former, by supplying the shadows: the same is evinced to the eye in the black and white of the definitive scale, Pl. 1. fig. 3. Perfect white is opaque, and perfect black transparent; hence, when added to black in minute proportion white gives it solidity; and from a like small proportion of black combined with white the latter acquires locality as a colour, and better preserves its hue in painting. Both white and black communicate these properties to other colours in proportion to their lightness or depth, while they cool each other in mixture, and equally contrast each other when opposed. These extremes of the chromatic scale are each in its way most easily defiled, as green, the mean of the scale, is the greatest defiler of colours. Rubens regarded white as the nourishment of light and the poison of shadow.

White is expressive of modesty and sweetness, and contributes to these expressions in other colours, when mixed therewith, by subduing their force; it is hence the pleasing expression of *paleness* and pureness of colour arises;—and in its general effect, as a colour on the eye and the mind, white is enlivening and elating, without gaiety, according to the neutrality of its relations; inspiring confidence or hope, as black or darkness does fear and distrust;—it has ever been the vesture of priesthood, and, in its sensible and moral expression, it is the natural garb and emblem of purity, delicacy, innocency, timidity, gentleness, dignity, piety, peace, and all the modest virtues; hence the *white flag* is the token of *peace*; the *white feather* the metaphor of *timidity;* and the *white vestments* of the vestal and priest the symbols of *purity* and *peace:* and it heightens these sentiments in pictorial representations, and lends its powers to language metaphorically; hence the poet also employs it ideally and rhetorically for all this variety of expression in the construction of epithets and the clothing of figures and symbols; and this he does likewise with all colours in the manner, reference, and relation, and with the same feeling as the painter.

Spenser, who was a great poetical colourist, gives this moral colouring to his figures; thus his Humbleness, as *Humilta,*

————Was an aged sire all *hoary gray*.
FAIRIE QUEEN, Cant. x. 5.

His Reverence,
Right cleanly clad in *comely sad* attire.
C. x. 7.

His Faith, as *Fidelia,*
She was arrayed all in *lily white*.
C. x. 13.

His Hope, as *Speranza,*
Was clad in *blue* that her beseemed well.
C. x. 14.

His Charity, as *Charissa,*
Was all in *yellow* robes arrayed.
C. x. 30.

His Falsehood,
Clad in *scarlet-red*
Purfled with *gold* and pearl of rich assay.
C. xi. 13.

His Praise-desire,
In a long *purple* pall, whose skirt with *gold*
Was fretted all about, she was arrayed.
C. ix. 37.

His Idleness,
——— The nurse of sin
Arrayed in habit *black* and amice thin.
C. iv. 18.

And many others. Indeed there is hardly a virtue, vice, or quality, which

Spenser has not figured and decorated with generally appropriate and expressive colours in his "Fairie Queen."

But of white, in particular, as employed by the poets, the following may serve as examples of epithets,—

White-robed truth.
MILTON.

White-robed innocence.
POPE.

Celestial *Piety!*
Mild in thy *milk-white* vest, to soothe my friend,
With holy fillets on thy brows descend.
ADDISON, after Statius, Silv. Lib. III.

The *saintly* veil of *maiden white.*
MILTON.

Vestures of *nuptial white.*
ROGERS.

O welcome *pure-eyed* Faith, *white-handed Hope,*
Thou hovering angel, girt with *golden* wings,
And thou *unblemish'd* form of *Chastity!*
MILTON'S COMUS.

In the latter example white is naturally and beautifully associated with yellow. White appears to be almost as principal in the colouring of the poet as of the painter, and volumes might be filled with instances of its use, but the following may suffice:—

White as the sunshine stream thro' *vernal clouds.*
AKENSIDE.

White as thy fame, and as thy honour *clear.*
DRYDEN.

Came vested all in *white, pure* as her mind.
MILTON.

Cytherea,
How bravely thou becom'st thy bed! *Fresh lily!*
And *whiter* than the sheets.
SHAKSPEARE'S CYMBELINE, Act II. Sc. 2.

—— While she, the picture of *pure piety,*
Like a *white* hind under the gripe's hard claws.
IBID.

No *whiter* page than Addison's remains.
POPE.

In the following, white is symbolical of pureness, innocence, chastity, candour, peace, friendship, and virtue in general.

> The *snowy* wings of *Innocence* and Love.
> AKENSIDE.

> Let *hoary* Judgment, sober guest,
> Bring *candour* in her *lilied* vest.
> MASON.

> ————Kind *Peace* appear,
> And in thy right hand hold the wheaten ear,
> From thy *white* lap th' o'erflowing fruits shall fall.

> She first, *white Peace*, the earth with plowshares broke,
> And bent the oxen to the crooked yoke.
> ADDISON AFTER TIBULLUS.

"You may find too the colour of the drapery that she [Friendship] wore in the old Roman paintings, from that verse in Horace,—

> Te *Spes*, et *albo* rara *Fides* colit
> Velata panno.   OD. 35. LIB. I."
> ON ANTIENT MEDALS, DIAL. I. p. 43.

> Now, by my *maiden honour*, yet as *pure*
> As the *unsullied lily*, I protest.
> SHAKSPEARE.

> To bring *white* hairs unto a *quiet* grave.
> IBID.

The following afford examples of white in accordance with red, of which relation we have already spoken, and shall have occasion to mention hereafter.

> Upon those lips, the sweet fresh buds of youth,
> The holy dew of prayer lies like a *pearl*
> Dropp'd from the opening eye-lids of the morn
> Upon the *bashful rose*.
> MIDDLETON.

Whose *blush* does thaw the *consecrated snow*
That lies on Dian's lap.
CANONS OF CRITICISM.

Ne less praiseworthy is her sister dear,
    Faire Mariane, the Muses' only darling;
Whose beautie *shineth as the morning clear*
    With *silver* dew upon the *roses pearling*.
SPENSER'S COL. CLO.

            What if this cursed hand
Were thicker than itself with *brother's blood?*
Is there not rain enough in the sweet heavens
To wash it *white as snow?*
HAMLET, Act III. Sc. 3.

To thee, sweet smiling maid, I bring
The beauteous progeny of spring;
In every *breathing bloom* I find
Some pleasing *emblem* of thy mind.
The *blushes* of that op'ning *rose*
Thy tender modesty disclose —.
The *snow-white lilies* of the vale,
Diffusing fragrance to the gale,
No ostentatious tints assume,
Vain of their exquisite perfume;
Careless, and sweet, and mild, we see
In them a *lovely type* of thee.
RICHARDSON'S RUSSIAN ANECDOTES.

——Yet I'll not shed her *blood;*
Nor scar that *whiter* skin of hers than *snow,*
And smooth as monumental *alabaster.*
OTHELLO, Act v. Sc. 2.

A very different form did *Virtue* wear;—
Rude from her forehead fell th' unplaited hair;
With dauntless mien aloft she rear'd her head,
And next to manly was the virgin's tread;
Her height, her sprightly *blush*, the goddess show,
And *robes unsullied as the falling snow.*
——————— She said ———————
With me the foremost place let Honour gain,
Fame and the praises mingling in her train;
Gay Glory next, and *Victory* on high,
White like myself on snowy wings shall fly.
ADDISON, AFTER SILIUS ITALICUS.

In the following, white is contrasted with black:—

>Whiter than new *snow* on a *raven's* back.
>
>ROMEO AND JULIET, Act III. Sc. 2.

>The Moon,
>Rising in *clouded* majesty, at length
>Apparent queen *unveil'd her peerless light,*
>And o'er the *dark* her *silver mantle* threw.
>
>MILTON.

>The Greeks obey; where yet the embers glow,
>Wide o'er the pile the *sable* wine they throw.
>Next the *white* bones his sad companions place,
>With tears collected, in a golden vase.
>
>POPE'S HOMER'S IL. B. XXIII.

>While, her *dark eyes* declining by his side,
>Moves in her *virgin veil* the gentle bride.
>
>ROGERS.

>*Dark*-wounding Calumny
>The *whitest* virtue strikes.
>
>SHAKSPEARE.

>Thou tremblest, and the *whiteness* on thy cheek
>Is apter than thy tongue to tell thy errand.
>E'en such a man, so *faint*, so *spiritless*,
>So *dull*, so *dead in look*, so *woe-begone*,
>Drew Priam's curtain in the *dead of night*.
>
>IBID.

In the latter of these passages white gives contrast or character to every thought, and heightens the whole sentiment, while in that which precedes it, it is contrasted and rendered more vivid. To conclude:—

>*White*, when it shines with unstain'd lustre clear,
>May bear an object back or bring it near.
>Aided by *black*, it to the front aspires;
>That aid withdrawn, it distantly retires;
>But black unmix'd, of darkest midnight hue,
>Still calls each object nearer to the view.
>
>MASON'S DUFRESNOY.

Of these poetical examples we have given more than might appear justifiable, because we believe, independent of the pleasure they may afford, the perusal of them will help to imbue the mind with just references and right appliances of colours.

White, as a pigment, is of more extensive use than any other colour in oil painting and fresco, owing to its local property, its representing light, and its entering into composition with all colours in forming tints: hence every artist is sensible of the importance of good pigments of this denomination; and Titian is said to have grieved, in one of his epistles, the death of the chemist who prepared his *white*, with the pathetic lamentation of a lover. The old masters are supposed to have possessed whites superior to our own; nevertheless we question this as a general fact, attributing the pureness of the local whites of some celebrated old pictures to faithful preparation,—a proper mode of using,—careful preservation of the work, and, in many instances, to the introduction of ultramarine, or a permanent cold colour into the white, helped also frequently by judicious contrast.

Notwithstanding white pigments are an exceedingly numerous class, an unexceptionable white is still a desideratum. The white earths are destitute of body in oil and varnish, and metallic whites of the best body are not permanent in water; yet when properly discriminated, we have eligible whites for most purposes, of which the following are the principal:—

I. WHITE LEAD, or ceruse, and other white oxides of lead, under the various denominations of *London* and *Nottingham whites*, &c. *Flake white*, *Crems* or *Cremnitz white*, *Roman* and *Venetian whites*, *Blanc d'argent* or *Silver white*, *Sulphate of lead*, &c. The heaviest and whitest of these are the best, and in point of colour and body are superior to all other whites. They are all, when pure and properly applied in oil and varnish, safe and durable, and dry well; but excess of oil discolours them, and in water-painting they are changeable even to blackness. They have also a destructive effect upon all vegetal lakes, except the rubial or madder lakes, and madder carmines; they are equally injurious to red and orange leads or minium, king's and patent yellow, massicot, gamboge, orpiments, &c.: but ultramarine, red and orange vermilions, yellow and orange chromes, madder colours, Sienna earth, Indian red, and all the ochres, compound with these whites with little or no injury. Cleanliness in using them is necessary for health; for though not virulently poisonous, they are pernicious when taken into or imbibed by the pores or otherwise, as are all other pigments of which lead is the basis. A fine natural white oxide, or carbonate of lead, would be a valuable acquisition.

The following are the true characters of these whites according to our particular experience:

1. LONDON and NOTTINGHAM WHITES. The best of these do not differ in any essential particular mutually, nor from the white leads of other manufactories. The latter, being prepared from flake white, is generally the grayest of the two. The inferior white leads are adulterated with whitening or other earths, which injure them in body and brightness, dispose them to dry more slowly, to keep their place less firmly, and to discolour the oil with which they are applied. All the above are carbonates of lead, and liable to froth or bubble when used with aqueous, spirituous, or acid preparations.

2. CREMS or CREMNITZ WHITE is a white carbonate of lead, which derives its names from Crems or Krems in Austria, or Kremnitz in Hungaria, and is called also *Vienna white*, being brought from Vienna in cakes of a cubical form. Though highly reputed, it has no superiority over the best English white leads, and varies like them according to the degrees of care or success with which it has been prepared.

3. FLAKE WHITE is an English white lead in form of scales or plates, sometimes gray on the surface. It takes its name from its figure, is equal or sometimes superior to Crems white, and is an oxidised carbonate of lead, not essentially differing from the best of the above. When levigated, it is also called *Body-white*.

4. BLANC D'ARGENT, or *Silver white*. These are false appellations of a white lead, called also *French white*. It is brought from Paris in the form of drops, is exquisitely white, and has all the properties of the best white leads; but, being liable to the same changes, is unfit for general use as a water-colour, though good in oil or varnish.

5. ROMAN WHITE is of the purest white colour, but differs from the former only in the warm flesh-colour of the external surface of the large square masses in which it is usually prepared.

6. SULPHATE OF LEAD is an exceedingly white precipitate from any solution of lead by sulphuric acid, much resembling the blanc d'argent; and has, when well prepared, quite neutral, and thoroughly edulcorated or

washed, most of the properties of the best white leads, but is sometimes rather inferior in body and permanence.

The above are the principal whites of lead; but there are many other whites used in painting, of which the following are the most worthy of attention:

II. ZINC WHITE is an oxide of zinc, which has been more celebrated as a pigment than used, being perfectly durable in water and oil, but wanting the body and brightness of fine white leads in oil; while in water, constant or barytic white, and pearl white, are superior to it in colour, and equal in durability. Nevertheless, zinc white is valuable, as far as its powers extend in painting, on account of its durability both in oil and water, and its innocence with regard to health.

III. TIN WHITE resembles zinc white in many respects, but dries badly, and has even less body and colour in oil, though superior to it in water. It is the basis of the best white in enamel painting.

There are various other *metallic whites*,—such are those of *bismuth, antimony, quicksilver*, and *arsenic;* but none of them are of any value or reputation in painting, on account of their great disposition to change of colour, both by light and foul air, in water and in oil.

IV. PEARL WHITE. There are two pigments of this denomination: one falsely so called, prepared from bismuth, which turns black in sulphureted hydrogen gas or any impure air, and is used as a cosmetic; the other, prepared from the waste of pearls and mother-of-pearl, which is exquisitely white, and of good body in water, but of little force in oil or varnish; it combines however with all other colours, without injuring the most delicate, and is itself perfectly permanent and innoxious;—witness Cleopatra's potation of pearls.

V. CONSTANT WHITE, PERMANENT WHITE, or *Barytic white*, is a sulphate of barytes, and when well prepared and free from acid, is one of our best whites for water-painting, being of superior body in water, but destitute of this quality in oil.

As it is of a poisonous nature, it must be kept from the mouth;—in other respects and properties it resembles the true pearl white. Both these pigments should be employed with as little gum as possible, as it destroys their body, opacity, or whiteness; and solution of gum ammoniac answers best.

VI. WHITE CHALK is a well-known native carbonate of lime, used by the artist only as a crayon, or for tracing his designs, for which purpose it is sawed into lengths suited to the port-crayon. *White crayons* and *tracing-chalks*, to be good, must work and cut free from grit. From this material *whitening* and lime are prepared, and are the bases of many common pigments and colours used in distemper, paper-staining, &c.

There are many other terrene whites under equivocal names, from the famed *Melinum*, or white earth of *Melos*, mentioned by Pliny to have been used by the Greek painters, to common whitening prepared from chalk. Among them are *Morat* or *Modan white*, *Spanish white*, or *Troys*, or *Troy white*, *Rouen white*, *Bougeval white*, *Paris white*, *Blanc de Roi*, *China white*, *Satin white*, &c. The common oyster-shell contains also a soft white in its thick part, which is good in water; and egg-shells have been prepared for the same purpose, as may likewise an endless variety of native earths, as well as those produced by art. From this unlimited variety of terrene whites we have selected above such only as are reputed, or as principally merit the attention of the artist;—the rest may be variously useful to the paper-stainer, plaisterer, and painter in distemper; but the whole of them are destitute of body in oil, and injurious to many colours in water, as they are to all colours which cannot be employed in fresco. See Tab. ix. Chap. xxii.

CHAP. IX.

OF THE PRIMARY COLOURS.

OF YELLOW.

>   What is here?
> *Gold? yellow, glittering, precious gold?* No, Gods,
> I am no idle votarist:———
> Thus much of this, will make *black white*, foul fair,
> Wrong right, base noble, old young, coward valiant.
> Ha! you Gods! why this———
> Will lug your priests and servants from your sides;
> Pluck stout men's pillows from below their heads;
> This *yellow* slave.
>   SHAKSP., TIMON OF ATHENS.

YELLOW is the first of the primary or simple colours, nearest in relation to, and partaking most of the nature of, the neutral white; it is accordingly a most advancing colour, of great power in reflecting light. Compounded with the primary *red*, it constitutes the secondary *orange*, and its relatives, scarlet, &c. and other warm colours.

It is the archeus or prime colour of the tertiary *citrine;*—it characterizes in like manner the endless variety of the semineutral colours called *brown*, and enters largely into the complex colours denominated buff, bay, tawny, tan, dan, dun, drab, chesnut, roan, sorrel, hazel, auburn, isabela, faun, feuillemorte, &c. Yellow is naturally associated with red in transient and prismatic colours, and they comport themselves with similar affinity and glowing accordance in painting, as well in conjunction as composition. It is the principal power also with red in representing the effects of warmth, heat, and fire, in painting and poetry:—

>   Where *Indian suns* engender new diseases,—
>   Where snakes and tigers breed,—I bend my way,
> To brave the *fev'rish thirst* no art appeases,—
>   The *yellow plagues*, and madd'ning *blaze of day*.
>   FROM THE SPANISH OF GONZALVO.

OF THE PRIMARY COLOURS.—OF YELLOW.

In combination on the other hand with the primary *blue*, yellow constitutes all the variety of the secondary *green*, and, subordinately, the tertiaries *russet* and *olive*. It enters also in a very subdued degree into cool, semi-neutral, and broken colours, and assists in minor proportion with blue and red in the composition of *black*.

As a pigment, yellow is a tender delicate colour, easily defiled, when pure, by other colours. In painting it diminishes the power of the eye by its action in a strong light, while itself becomes less distinct as a colour; and, on the contrary, it assists vision and becomes more distinct as a colour in a neutral somewhat declining light. These powers of colours upon vision require the particular attention of the colourist. To remedy the ill effect arising from the eyes having dwelt upon a colour, they should be gradually passed to its opposite colour, and refreshed amid compound or neutral tints, or washed in the clear light of day.

In a warm light, yellow becomes totally lost, but is less diminished than all other colours, except white, by distance. The stronger tones of any colour subdue its fainter hues in the same proportion as opposite colours and contrasts exalt them. The contrasting colours of yellow are a purple inclining to blue when the yellow inclines to orange, and a purple inclining to red when the yellow inclines to green, in the mean proportions of *thirteen* purple to *three* of yellow, measured in surface or intensity; and yellow being nearest to neutral white in the natural scale of colours it accords with it in conjunction. Of all colours, except white, it contrasts black most powerfully.

Yellow is discordant when standing alone, or unsupported by other colours, with orange. It is the vulgar symbol of jealousy, occasioned perhaps by the biliary complexion attending that passion; to which symbol Butler alludes thus:—

> *Jealous* piques,
> Which th' antients wisely signified
> By th' *yellow* manto's of the bride.
> HUDIBR., Part II. Canto I.

And Chaucer thus:—

> And *Jalousie*,
> That wered of *yelw colors* a gerlond
> And had a cuckow sitting on hir hond.
> KNIGHT'S TALE, v. 1032.

yet the sensible effects of yellow are gay, gaudy, glorious, full of lustre, enlivening, and irritating; and its impressions on the mind partake of these characters, and acknowledge also its discordances:—

> Only they
> That come to hear a merry, *gaudy play*,
> A noise of targets; or to see a fellow
> *In a long motley coat, guarded with yellow*,
> Will be deceived.
>
> SHAKSP., PROL. HEN. VIII.

The name yellow is used metaphorically for several malign passions; but from want of euphony or other cause, it is less employed than those of other colours by the poets, with whom the terms saffron, golden, orient, &c. supply its place; hence—

> Now when the rosy morn began to rise,
> And wave her *saffron* streamer.
>
> DRYDEN.

> Now when the *rosy-finger'd* morning *faire*,
> Weried of aged Tithon's *saffron* bed,
> Had spred her *purple* robe through dewy aire.
>
> FAIRIE QUEEN, Cant. II. 7.

> The cynosure of *jaundiced* eyes.
>
> SHAKSPEARE.

> Soon as the white and red mixt finger'd dame
> Had *gilt* the mountain with her *saffron* flame.
>
> CHAPMAN'S ODYSSEY.

The substitution of *gold*, &c. for *yellow* by the poets may have arisen not less from the great value and splendour of the metal, than from the paucity of fine yellows among those antients who celebrated the Tyrian purple or red, and the no less famed Armenian blue;—so in the beautiful illuminated Mss. of old, and in many antient paintings, which glowed with vermilion and ultramarine, the place of yellow was supplied by gilding:* had there been blue and red metals equal in beauty to the yellow of gold, their brilliancy would probably have driven other coloured pigments from the field of early art; but the modesty of nature has wisely denied such meretricious beauty to the painter; metallic tones being as harsh and unsavoury to chaste sense in painting as they are in music.

* This is also remarkable in the unique collection of antient oil paintings belonging to Charles Aders, Esq.

OF THE PRIMARY COLOURS.—OF YELLOW. 75

In the next example, the poet gives force to his epithets and comparisons by a double contrast, equally true and beautiful :—

> So sweet a kiss the *golden sun* gives not
> To those fresh morning drops upon the *rose;*
> Nor shines the *silver moon* one half so bright
> Through the transparent bosom of the deep.
>
> SHAKSPEARE.

In the following passage he employs yellow judiciously and naturally in various of its relations, and invigorates them by purple in the midst, as in the preceding instance from Spenser's Fairie Queen :—

> Your *straw-coloured* beard, your *orange-tawney* beard, your *purple-in-grain* beard, or your French *crown-coloured,* your *perfect yellow.*

In the following, Shakspeare characterizes yellow metaphorically for jealousy :—

> I will possess him with *yellowness,*
> For the revolt of mien is dangerous.

In the next, as giving lustre and life :—

> Glittering in *golden* coats, like images,
> As full of spirits as the month of May,
> And *gorgeous as the sun at midsummer.*
>
> IDEM.

Here, as irritating :—

> He will come to her in *yellow* stockings,
> And 'tis *a colour she abhors.*
>
> IDEM.

Again, variously :—

> Two beauteous springs to *yellow autumn* turn'd.
>
> IDEM.

> Princes,
> What grief hath set the *jaundice* on your cheeks?
>
> IDEM, TROIL. AND CRESSID. A. I. Sc. 3.

> Do not, as some ungracious pastors do,
> Show me the steep and thorny way to heaven ;
> Whilst, like a puff'd and reckless libertine,
> Himself the *primrose* path of dalliance treads.
>
> IDEM, HAMLET, A. I. Sc. 3.

Now he shades it:—

> Her hair is *auburn*, mine is perfect *yellow*;
> If that be all the difference in his love,
> I'll get me such a colour'd periwig.
> SHAKSPEARE.

> I have lived long enough: my way of life
> Is fall'n into the *sear*, the *yellow leaf*.
> IDEM.

And here he contrasts it with black:—

> Not *black* in my mind, though *yellow* in my legs.
> IDEM.

> Or hide me *nightly* in a charnel-house,
> O'er-cover'd quite with dead men's rattling bones,
> With reeky shanks, and *yellow* chapless skulls.
> ROMEO AND JULIET, A. IV. Sc. 1.

> On his harnais,—
> With nailes *yelwe* and bright as any *gold*,
> He had a beres skin, *cole-blake* for old.
> His long haire was kempt behind his bak:
> As any ravene's fethere it shone for *blake*.
> CHAUCER'S KNIGHT'S TALE, v. 2142.

By other poets the term yellow is almost universally substituted metonymously, as we have already instanced; yet Spenser sings:—

> Her *yellowe* locks, that shone so bright and long,
> As sunny beams in fairest summer's day;
> She fiercely tore, and with outrageous wrong
> From her *red* cheeks the *roses* rent away.
> COL. CLO.

Goldsmith, another of nature's pupils, has celebrated—

> The *yellow-blossom'd* vale;

and Byron, in imitating Catullus, speaks of—

> The *yellow harvest's* countless seed.

Yellow is a colour abundant throughout nature, and its class of pigments abounds in similar proportion. We have arranged them under the following heads, agreeably to our plan, according to their definiteness and brilliancy of colour; first, the opaque, and then the transparent, or finishing colours. It may be observed of yellow pigments, that they much resemble whites in their chemical relations in general.

I. 1. CHROME YELLOW is a pigment of modern introduction into general use, and of considerable variety, which are mostly *chromates of lead*, in which the latter metal more or less abounds. They are distinguished by the pureness, beauty, and brilliancy of their colours, which qualities are great temptations to their use in the hands of the painter; they are notwithstanding far from unexceptionable pigments;—yet they have a good body, and go cordially into tint with white both in water and oil; but, used alone or in tint, they after some time lose their pure colour, and may even become black in impure air; they nevertheless resist the sun's rays during a long time. Upon several colours they produce serious changes, ultimately destroying Prussian and Antwerp blues, when used therewith in the composition of greens, &c. In general they do not accord with the modest hues of nature, nor harmonize well with the sober beauty of other colours; hence the opinions of artists vary exceedingly respecting these pigments. Whether improvement in their modes of preparation and use will render them eligible pigments hereafter remains yet to be proved.

We have prepared them upon almost every possible base; and the late ingenious Dr. Bollmann, who introduced them into commerce, made trials at our suggestion for improving them, but none has been hitherto produced upon which the artist can safely trust his reputation as a colourist. This substance was known as a native pigment long before it was distinguished as a chemical substance.

2. JAUNE MINERALE. This pigment is also a chromate of lead, prepared in Paris, differing in no essential particular from the above, except in the paleness of its colour. The chrome yellows have also obtained other names from places or persons from whence they have been brought, or by whom they have been prepared; we pass over, however, such as have not been generally received. The following pigment passes also under the name of Jaune Minerale:—

II. PATENT YELLOW, *Turner's Yellow*, or *Montpellier Yellow*, is a submuriate or cloruret of lead, which metal is the basis of most opaque yellow pigments: it is a hard, ponderous, sparkling substance, of a crystalline texture and bright yellow colour; hardly inferior, when ground, to chromic yellow. It has an excellent body, and works well in oil and water, but is soon injured both by the sun's light and impure air; it is therefore little used, except for the common purposes of house-painting, &c.

III. TURBITH MINERAL is a sub-sulphate of mercury, of a beautiful lemon yellow colour, but so liable to change by the action of light or impure air, that, notwithstanding it has been sometimes employed, it cannot be used safely, and hardly deserves attention as a pigment.

IV. NAPLES YELLOW is a compound of the oxides of lead and antimony, antiently prepared at Naples under the name of *Giallolini;* it is supposed also to have been a native production of Vesuvius and other volcanoes, and is a pigment of deservedly considerable reputation. It is not so vivid a colour as either of the above, but is variously of a pleasing light, warm, yellow tint. Like all the preceding yellows it is opaque, and in this sense is of good body. It is not changed by the light of the sun, and may be used safely in oil or varnish, under the same management as the whites of lead; but, like these latter pigments also, it is liable to change even to blackness by damp and impure air when used as a water-colour, or unprotected by oil or varnish.

Iron is also destructive of the colour of Naples yellow, on which account great care is requisite in grinding and using it not to touch it with the common steel palette-knife, but to compound its tints on the palette with a spatula of ivory or horn. For the same reason it may be liable to change in composition with the ochres, Prussian and Antwerp blues, and all other pigments of which iron is an ingredient or principal; but used purely or with white lead, its affinity with which gives permanency to their tints, it is a valuable and proved colour in oil, in which also it works and dries well.

It may also be used in enamel painting, as it vitrifies without change, and in this state it was formerly employed under the name of *Giallolini di fornace*, and has been again introduced, under an erroneous conception that vitrification gives permanence to colours, when in truth it only increases the difficulty of levigation, and injures their texture for working. Naples yellow does not appear to have been generally employed by the early painters in oil.

V. ANTIMONY YELLOW is also a preparation of antimony, of a deeper colour than Naples yellow, and similar in its properties. It is principally used in enamel and porcelain painting.

VI. MASSICOT, or *Masticot*, is an oxide of lead of a pale yellow colour, exceedingly varying in tint from the purest and most tender yellow or straw colour to pale ash colour or gray. It has in painting all the properties of the white lead from which it is prepared, but in tint with which nevertheless it soon loses its colour and returns to white:—if, however, it be used pure or unmixed, it is a useful delicate colour, permanent in oil under the same conditions as white lead, but ought not to be employed in water, on account of its changing in colour even to blackness by the action of damp and impure air. It appears to have been prepared with great care, and successfully employed by the old masters.

VII. YELLOW OCHRE, called also *Mineral Yellow*, is a native pigment, found in most countries, and abundantly in our own. It varies considerably in constitution and colour, in which latter particular it is found from a bright but not very vivid yellow to a brown yellow, called *spruce ochre*, and is always of a warm cast. Its natural variety is much increased by artificial dressing and compounding. The best yellow ochres are not powerful, but as far as they go are valuable pigments, particularly in fresco and distemper, being neither subject to change by ordinary light, nor much affected by impure air or the action of lime; by time, however, and the direct rays of the sun they are somewhat darkened. Iron is the principal colouring matter in them all, of which the following are the principal species; but they are often confounded:—

1. OXFORD OCHRE is a native pigment from the neighbourhood of Oxford, semi-opaque, of a warm yellow colour and soft argillaceous texture, absorbent of water and oil, in both which it may be used with safety according to the general character of yellow ochres, of which it is one of the best. Similar ochres are found in the Isle of Wight, in the neighbourhood of Bordeaux, and various other places.

2. STONE OCHRE has been confounded with the above, which it frequently resembles, as it does also Roman ochre. True stone ochres are found in balls or globular masses of various sizes in the solid body of stones, lying near the surface of rocks among the quarries in Gloucestershire and

elsewhere. These balls are of a smooth compact texture, in general free from grit, and of a powdery fracture. They vary exceedingly in colour from yellow to brown, murrey and gray, but do not differ in other respects from the preceding, and may be safely used in oil or water in the several modes of painting, and for browns and dull reds in enamel.

3. DI PALITO is a light yellow ochre, not differing much from the foregoing, but affording tints rather purer in colour than the brightest of them, but less so than Naples yellow. Many pleasing varieties of ochrous colours are produced by burning and compounding with lighter, brighter, and darker colours, but often very injudiciously, and adversely to that simple economy of the palette which is favourable to the certainty of operation, effect, and durability.

4. ROMAN OCHRE is rather deeper and more powerful in colour than the above, but in other respects differs not essentially from them;— a remark which applies equally to yellow ochres of other denominations.

5. BROWN OCHRE, *Spruce Ochre*, or *Ochre de Rue*, is a dark-coloured yellow ochre, in no other respect differing from the preceding :—it is much employed, and affords useful and permanent tints.

VIII. TERRA DI SIENNA, or *Raw-Sienna earth*, &c. is also a ferruginous native pigment, and appears to be an iron ore, which may be considered as a crude natural yellow lake, firm in substance, of a glossy fracture, and very absorbent. It is in many respects a valuable pigment,—of rather an impure yellow colour, but has more body and transparency than the ochres; and being little liable to change by the action of either light, time, or impure air, it may be safely used according to its powers, either in oil or water, and in all the modes of practice.

IX. 1. YELLOW ORPIMENT is a sulphurated oxide of arsenic, of a beautiful, bright, and pure yellow colour, not extremely durable in water, and less so in oil :—in tint with white lead it is soon destroyed;—it is not subject to discolourment in impure air :—this property is not, however, sufficient to redeem it with the artist, as it has a bad effect upon several valuable colours; and, although it is not so poisonous as white arsenic, is dangerous in its effect upon health. Yellow orpiment is of several tints, from bright cool yellow to warm orange, the first of which are most subject to change; and it has appeared under various forms and denominations :—these seem to have been used by several of the old masters, with especial care to avoid mixture; and as they dry badly, and the oxides of lead used in rendering oils drying destroy their colour, levigated glass was employed with them as a dryer, or perhaps they were sometimes used in simple varnish.

We know that Mengs and Sir Joshua Reynolds employed them in their practice, as did also Snyders, John Van Huysum, De Heem, and other painters of still life, sometimes successfully, and sometimes otherwise; but we are far from recommending them as eligible pigments.

2. KING'S YELLOW. Yellow orpiment has been much celebrated under this name, as it has also under the denomination of—

3. CHINESE YELLOW, which is a very bright sulphuret of arsenic brought from China.

X. 1. PLATINA YELLOW is, as its name implies, a preparation from platina, which has afforded the author two fine yellow pigments, the deepest of which resembles the Terra di Sienna (VIII.), but is warmer in tone and richer in colour and transparency, much resembling fine gall-stone. It works well, and is extremely permanent both in water and oil, in which it neither suffers change by the action of the sun, nor of sulphuretted hydrogen and impure air, and is therefore a valuable pigment, and may be produced of many tints. The other is a—

2. LEMON YELLOW, of a beautiful light vivid colour. In body and opacity it is nearly equal to Naples yellow and masticot, but much more pure and lucid in colour and tint, and at the same time not liable to change by damp, sulphureous or impure air, nor by the action of light, nor by the steel palette-knife, nor by mixture with white lead or other pigments, either in water or oil, in each of which vehicles it works well. Both these pigments are, therefore, valuable additions to the palette. Lemon yellow is principally adapted to high lights in painting. In water it exceeds gamboge in brightness, and in mixture therewith improves its beauty. This mixture also goes readily into oil: indeed it is the best and easiest way of rendering gamboge diffusible as an oil-colour,—simple solution of the gamboge in little water, and trituration of the lemon-yellow therewith, being all that is requisite for this purpose. The light yellow of the definitive scale, Pl. I. fig. 3. is of this pigment, which, being unaffected by lime, is eligible also in fresco and crayons.

XI. MADDER YELLOW is a preparation from the madder-root. The best is of a bright colour, resembling Indian yellow, but more powerful and transparent, though hardly equal to it in durability of hue,—metallic, terrene, and alkaline substances acting on and reddening it as they do gamboge: even alone it has by time a natural tendency to become orange and foxy. We have produced it of various hues and tints, from an opaque and ochrous yellow, to a colour the most brilliant, transparent, and deep. Upon the whole, however, after an experience of many years, we do not consider them eligible pigments.

XII. 1. GAMBOGE; or, as it is variously written, Gumboge, Camboge, Gambouge, Cambogia, Cambadium, Cambogium, Gambadium, Gambogium, &c. is brought principally, it is said, from Cambaja in India, and is, we are told, the produce of several trees. The natives of the coast of Coromandel call the tree from which it is principally obtained *Gokathu*, which grows also in Ceylon and Siam. From the wounded leaves and young shoots the gamboge is collected in a liquid state, and dried: indeed our indigenous herb celandine yields, in the same manner, a beautiful yellow juice of the same properties as gamboge. Gamboge is a concrete vegetal substance, of a gum-resinous nature, and beautiful yellow colour, bright and transparent, but not of great depth. When properly used, it is more durable than generally reputed both in water and oil: it is deepened in some degree by ammoniacal and impure air, and somewhat weakened, but not easily discoloured, by the action of light. Time effects less change in this colour than on other bright vegetal yellows; but white lead and other metallines injure, and terrene and alkaline substances redden it. It works remarkably well in water, with which it forms an opaque solution, without grinding or preparation, by means of its natural gum; but is with difficulty used in oil, &c. in a dry state. In its natural state it however dries well, and lasts in glazing when deprived of its gum. It is perfectly innocent with regard to other colours, and, though it is a strong medicine, is not dangerous or deleterious in use. Sir Joshua Reynolds and Wilson are said to have employed it, and so also we know did the amiable President, West: the first of these used it softened into a paste with water, and the latter in a dry state precipitated upon whitening. It has also been employed as a yellow lake prepared upon an aluminous base; but a much better way than either is to dissolve it into a paste in water, and mix it with lemon-yellow, with which pigment being diffused it goes readily into oil or varnish.

2. EXTRACT OF GAMBOGE is the colouring matter of gamboge separated from its greenish gum and impurities by solution in alcohol and precipitation, by which means it acquires a powdery texture, rendering it miscible in oil, &c. and capable of use in glazing. It is at the same time improved in colour, and retains its original property of working well in water with gum.

XIII. GALL-STONE is an animal calculus formed in the gall bladder, principally of oxen. This concretion varies a little in colour, but is in general of a beautiful golden yellow, more powerful than gamboge, and is highly reputed as a water-colour: nevertheless, its colour is soon changed and destroyed by strong light, though not subject to alteration by impure air.

It is rarely introduced in oil painting, and is by no means eligible therein; and, as a water-colour, is every way inferior to platina yellow.

XIV. INDIAN YELLOW, brought from India, as its name implies, is a pigment long employed in India under the name *Pwree*, but has not many years been introduced generally into painting in Europe. It is imported in the form of balls, is of a fetid odour, and is produced from the urine of the camel. It has also been ascribed, in like manner, to the buffalo, or Indian cow, after feeding on mangos; but the latter statement is incorrect. However produced, it appears to be a urio-phosphate of lime, of a beautiful pure yellow colour, and light powdery texture; of greater body and depth than gamboge, but inferior in these respects to gall-stone. Indian yellow resists the sun's rays with singular power in water-painting; yet in ordinary light and air, or even in a book or portfolio, the beauty of its colour is not lasting. It is not injured by foul air, but in oil it is exceedingly fugitive, both alone and in tint. Owing probably to its alkaline nature, it has injurious effect upon cochineal lakes and carmine when used with them. As lime does not injure this colour, it may be employed in fresco.

XV. 1. YELLOW LAKE. There are several pigments of this denomination, varying in colour and appearance according to the colouring substances used and modes of preparation. They are usually in the form of drops, and their colours are in general bright yellow, very transparent, and not liable to change in an impure atmosphere; qualities which would render them very valuable pigments, were they not soon discoloured, and even destroyed, by the opposite influence of oxygen and light, both in water and oil; in which latter vehicle, like other lakes in general, they are bad dryers, and do not stand the action of white lead or metallic colours. If used, therefore, it should be as simple as possible. Of these lakes, the following are the best :—

2. QUERCITRON LAKE, or *Quercitron Yellow*, is what its names imply. It is dark in substance, in grains of a glossy fracture, perfectly transparent, and when ground is of a beautiful yellow colour, more durable than the common yellow lakes, although not perfectly permanent.

In painting it follows and adds richness and depth to gamboge in water, and goes well into varnish; but the lead used in rendering oils desiccative, browns it, and for the same reason it is useless in tints.

XVI. DUTCH PINK, ENGLISH and ITALIAN PINKS, are sufficiently absurd names of yellow colours prepared by dyeing, whitening, &c. with vegetal yellow tinctures, in the manner of rose pink, from which they borrow their name.

They are bright yellow colours, extensively used in distemper and for paper-staining, and other ordinary purposes; but are little deserving attention in the higher walks of art, being in every respect inferior even to the yellow lakes, except the best kind of Italian pink, which is, in fact, a yellow lake, and richer in colour than the pigments generally called yellow lake.

The pigment called *Stil*, or *Stil de grain*, is a similar preparation, and a very fugitive yellow, the darker kind of which is called brown-pink.

CHAP. X.

OF RED.

Celestial *rosy red*, *Love's* proper hue.
MILTON'S PARADISE LOST.

RED is the second and intermediate of the primary colours standing between *yellow* and *blue;* and in like intermediate relation also to *white* and *black*, or light and shade. Hence it is pre-eminent among colours, as well as the most positive of all, forming with yellow the secondary *orange* and its near relatives, scarlet, &c.; and with blue, the secondary *purple*, and its allies, crimson, &c. It gives some degree of warmth to all colours, but most so to those which partake of yellow.

It is the archeus, or principal colour, in the tertiary *russet;* enters subordinately into the two other tertiaries, *citrine* and *olive;* goes largely into the composition of the various hues and shades of the semineutral *marrone*, or chocolate, and its relatives, puce, murrey, morello, mordore, pompadour, &c.; and more or less into *browns, grays*, and all broken colours. It is also the second power in harmonizing other colours, and in compounding *black* and all neutrals, into which it enters in the proportion of five, to blue eight, and yellow three.

Red is a colour of double power in this respect also; that, in union or connexion with yellow, it becomes hot and *advancing;* but mixed or combined with blue, it becomes cool and *retiring*. It is, however, more congenial with yellow than with blue, and thence partakes more of the character of the former in its effects of warmth, of the influence of light and distance, and of action on the eye, by which the power of vision is diminished upon viewing this colour in a strong light; while, on the other hand, red itself appears to deepen in colour rapidly in a declining light as night comes on, or in shade. These qualities of red give it great importance, render it difficult of management, and require it to be kept in general subordinate in painting: hence it is rarely used unbroken, or as the archeus, ruling or predominating colour, or for toneing a picture; on which account it will always appear detached or insulated, unless it be repeated and subordinated in a composition. Accordingly Nature uses red sparingly, and with as great reserve in the decoration of her works as she is profuse in lavishing green upon them; which is of all colours the most soothing to the eye, and the true compensating colour, or contrasting or harmonizing equivalent of red, in the proportional quantity of eleven to five of red, according to surface or intensity; and is, when the red inclines to scarlet or orange, a *blue-green;* and, when it inclines to crimson or purple, is a *yellow-green.*

Red breaks and diffuses with white with peculiar loveliness and beauty; but it is discordant when standing with orange only, and requires to be joined or accompanied by their proper contrast, to resolve or harmonize their dissonance.

In landscapes, &c. abounding with hues allied to green, a red object, properly posited according to such hues in light, shade, or distance, conduces wonderfully to the life, beauty, harmony, and connexion of the colouring.

Red being the most *positive* of colours, and having the *middle* station of the primaries, while *black* and *white* are the *negative* powers or neutrals of colours, and the *extremes* of the scale,—red contrasts and harmonizes these neutrals; and, as it is more nearly allied to white or light than to black or shade, this harmony is most remarkable in the union or opposition of white and red, and this contrast most powerful in black and red.

As a colour, red is in itself pre-eminently beautiful, powerful, cheering, splendid, and ostentatious, and communicates these qualities to its two secondaries, and their sentiments to the mind. Hence the blind man mentioned by Locke, who compared *scarlet* to the sound of a trumpet, had not a very bad conception of its analogy; nor very different from that of Euripides, where he says—

> But when the *flaming torch* was hurl'd, the sign
> Of *purple fight,* as when the *trumpet sounds,* &c.

and the intelligent youth couched by Cheselden thought scarlet the most beautiful of all colours. This beauty of red is a great temptation to that undue use of it which subjected the Greek painter Parrhasius to the censure of Euphranor, " that the Theseus of the former had eaten roses, but that his own had fed upon beef."

> The bold Erechtheus' son,
> Whom Pallas bred, and cherish'd as her own.
> PLUTARCH, ON THE FAME OF THE ATHENIANS.

The same beautiful fault has been attributed in modern times to the colouring of Rubens, whose very name may be supposed in a future age to have been given him on this account. To see things or events *en couleur de rose*, is, with our vivacious neighbours, to look upon them with a cheerful, partial, and favourable eye.

Nature, every where replete with benevolent intelligence and good taste, has given this colour to the blood, with the property of becoming more vivid when spilt; and has proportioned its action on the eye to the danger of the case, as when an artery is wounded. If it excite salutary terror by association, its immediate effect on the sense is to moderate alarm by its beauty. Had it been black instead of red, it would have been less promptly discovered, and more frightful in its appearance. But this is out of place, except as an instance that nature sanctions it as a true principle of art that we should disguise or veil whatever is horrible, abominable, or loathsome.

Red is expressive of ardour and the sanguine passions: it is hence peculiarly a military colour, as appropriate to war as white is to peace: hence the red plumes worn by military heroes in antient times. It dyes the flag of defiance, and is the emblem of blood; naturally stimulating and indicating fierceness and courage, as in the comb of the cock, the throttles of the turkey, and in exciting the bull to rage.

As powerful, it has become the symbol of power and distinction, and hence has decorated equally the regal robe and the mantle of martyrdom, producing awe, veneration, and fear; while in its gentler offices it moves and assists the affections of

> Love, Hope, and Joy, fair Pleasure's smiling train;—

and is, upon the whole, the most effective of colours. The poets have accordingly availed themselves freely of this colour and its progeny, for the purpose of expression, in decorating figures and constructing epithets, often using the term purple metonymically for red;—sometimes, it is true, for the mere words as sounds, but frequently also with the refined taste, true judgment, and cultivated feeling of the painter; of which, and as illustrations of the relations, attributes, and uses of this colour, the following examples may suffice, in reference to

1. Beauty, &c.

> *Beauty's ensign* yet
> Is *crimson* in thy lips and in thy cheeks,
> And Death's *pale flag* is not advanced there.
> SHAKSPEARE.

> To *blush* and *beautify* the cheek again.
> IDEM.

2. Joy, &c.

> See, your guests approach:
> Address yourself to entertain them sprightly,
> And let's be *red* with *mirth*.
> IDEM.

> They call drinking deep, dying *scarlet*.
> IDEM.

> *Blooming youth and gay delight*
> Sit on thy *rosy* cheeks confess'd.
> PRIOR.

3. Love, &c.

> Would you know where young *Love* in his *beauty* reposes,
>   And find him uninjured by sorrow or care,
> Go—seek for the boy in the *Valley of Roses*,
>   And find the pure spirit of constancy there.
> MARY ANN BROWN.

> Coarse complexions,
> And cheeks of ev'ry grain, will serve to ply
> The sampler, and to tease the housewife's wool:
> What need a *vermil-tinctured lip* for that—
> *Love-darting eyes*, or *tresses like the morn?*
> MILTON.

4. Hope, &c.

> For me the balm shall *bleed*, the amber flow,
> The *coral redden*, and the *ruby glow*.
> POPE.

> The *rosy*-finger'd morning *fair*.
> SPENSER.

5. Dignity, &c.

The *scarlet honour* of your peaceful gown.
DRYDEN.

Thy *ambition*,
Thou *scarlet* sin, robb'd this bewailing land
Of *noble* Buckingham.
SHAKSPEARE.

6. Ardour, &c.

He spoke: the goddess with the charming eyes
Glows with *celestial red*, and thus replies—.
POPE'S HOMER.

While Mars, descending from his *crimson* car,
Fans with fierce hands the kindling flames of war.
HALLER.

7. Anger, &c.

If I prove honey-mouth'd, let my tongue blister,
And never to my *red-look'd anger* be
The trumpet any more.
SHAKSPEARE.

How *bloodily* the sun begins to peer
Above yon busky hill! The day looks *pale*
At his distemperature.
IDEM.

Change the complexion of her *maid-pale peace*
To *scarlet indignation*, and bedew
Her *pastures' grass* with faithful English *blood*.
IDEM.

Spreads the *red* rod of *angry* pestilence.
MILTON.

8. Accordance with white, &c.

She gathereth flowers partie *white* and *rede*,
To make a sotel garland for hir head.
CHAUCER'S KNIGHT'S TALE.

Hath *white* and *red* in it such wondrous power,
That it can pierce through eyes into the heart,
And therein stir such rage and restless tower
As only death can stint the dolorous smart?
SPENSER'S HYMN TO BEAUTY.

FROM THE PERSIAN OF HAFIZ.

Sweet maid, if thou would'st *charm my sight*,
   And bid these arms thy neck infold;
     That *rosy* cheek, that *lily* hand,
Would give thy poet more delight
   Than all Bocara's vaunted gold,
     Than all the gems of Samarcand.

Boy, let you *liquid ruby* flow,
   And bid thy pensive heart be glad,
     Whate'er the frowning zealots say;
Tell them their Eden cannot show
   A stream so clear as Roenabad,
     A bower so sweet as Mosellay.
SIR W. JONES.

'Tis *beauty truly blent, whose red and white*
Nature's own sweet and cunning hand laid on.
SHAKSPEARE.

Such war of *white* and *red* within her cheek.
IDEM.

         Through whose *white* skin
With *damaske* eyes the *ruby blood* doth peep.
MARLOWE.

How now, my love? Why is your cheek so *pale?*
How chance the *roses* there do fade so fast?
SHAKSPEARE.

The seasons alter: *hoary-headed* frosts
Fall in the fresh lap of the *crimson rose.*
IDEM.

A pudency so *rosy*, the sweet view on 't
Might well have *warm'd* old Saturn; that I thought her
As *chaste as unsunn'd snow.*
IDEM, CYMBELINE, Act II. Scene 5.

  So women, to surprise us, spread
  Their borrow'd flags of *white* and *red.*
BUTLER, HUDIBRAS, Part II. Canto III.

Unto the ground she cast her modest eye,
  And ever and anon with *rosie red*
The *bashful blood* her *snowy* cheeks did dye,
That her became, as polish'd *ivory*
  Which cunning craftsman's hand hath overlaid
With fair *vermilion*, or pure lastery.
FAIRIE QUEEN, Canto IX. 41.

There grew a goodly tree him fair beside,
  Loaden with fruit and apples *rosie red*,
As they in pure *vermilion* had been dyed.
IDEM, Canto XI. 46.

O thus ... lay the gentle babes.
Thus ... girdling one another
Within their *alabaster innocent* arms,
Their lips were four *red roses* on a stalk,
Which, in their summer beauty, kiss'd each other.
RICHARD THE THIRD, Act IV. Scene 3.

9. Harmony with light, &c.

                    Morn,
Waked by the circling Hours, with *rosy* hand
Unbarr'd the gates of *light*.
MILTON.

And bid the cheek be ready with a *blush*
Modest as Morning, when she *coldly* eyes
The youthful Phœbus.
TROILUS AND CRESSIDA, Act I. Scene 3.

Say, that she frowns; I'll say, she looks as *clear*
As morning *roses* newly wash'd with dew.
SHAKSPEARE.

Of Nature's gifts thou mayst with *lilies* boast,
And with the *half-blown rose*.
IDEM.

Patience, thou young and *rose-lipp'd* cherubim.
IDEM, OTHELLO, Act IV. Scene 2.

  Here the *roses blush* so rare,
  Here the morning smiles so *fair*.
CRASHAW.

10. Contrast with harmony, &c.—

>See where she sits upon the *grassie green*
>   (O seemly sight!)
>Yclad in *scarlet* like a maiden queen,
>   And *ermines white.*
>Upon her head a *crimosin* coronet
>With *damask roses* and daffodils set,
>   *Bay leaves* between
>   And *primroses green,*
>Embellish the sweet violet.
>Tell me have ye seen her angel-like face,
>   Like Phœbe fair?
>Her heavenly 'haviour, her princely grace,
>   Can you well compare?
>The *red-rose* melted *with the white yfere,*
>In either cheek depeincten lively cheere:
>   Her modest eye,
>   Her majesty,
>Where have you seen the like but there?
>   SPENSER'S SHEP. CAL. AP.

>His hand did quake
>And tremble like an *aspin green;*
>And troubled *blood* through his *pale* face was seen
>To come and go; with tydings from the heart,
>As it a running messenger had been.
>   FAIRIE QUEEN, C. IX. 51.

>On her left breast,
>A mole cinque-spotted, like the *crimson* drops
>I' the bottom of a *cowslip.*
>   SHAKSPEARE, CYMBELINE, Act II. Sc. 2.

>And I serve the Fairy Queen,
>To dew her orbs upon the green:
>The *cowslips* tall her pensioners be;
>In their *gold coats* spots you see:
>Those be *rubies,* fairy favours;
>In those freckles live their savours:—
>I must go seek some dew-drops here,
>And hang a *pearl* in every cowslip's ear.
>   MID. NIGHT'S DREAM, Act II. Sc. 1.

11. Contrast with black, &c.—

>The air hath starved the *roses* in her cheeks,
>And pinch'd the *lily-tincture* of her face,
>That now she is become *as black as* I.
>   SHAKSPEARE.

Beaufort's *red sparkling* eyes blab his heart's malice,
And Suffolk's *cloudy brow* his stormy hate.
                                                    SHAKSPEARE.
And he to deck her *raven* hair
Would weave the *rose* and *lily fair*.
                                                    MARY A. BROWN.

Red is so much the instrument of beauty in nature and art in the colour of flesh, flowers, &c. that good pigments of this genus may of all colours be considered the most indispensable: we have happily, therefore, many of this denomination, of which the following are the principal:—

I. VERMILION is a sulphuret of mercury, which, previous to its being levigated, is called *cinnabar*. It is an antient pigment, the κιννάβαρι of the Greeks, and is both found in a native state and produced artificially. Vermilion probably obtained its name from resemblance, or admixture with the beautiful though fugitive colours obtained from the *vermes*, or insects, which yield carmines. [See Kermes Lake.] The Chinese possess a native cinnabar so pure as to require grinding only to become very perfect vermilion, not at all differing from that imported in large quantities from China: it is said also to be found in abundance in Corinthia, in the Palatinate, Friuli, Bohemia, Almaden in Spain, the principality of Deux-Ponts, and also in South America, particularly in Peru, &c.

Chinese vermilion is of a cooler or more crimson tone than that generally manufactured from factitious cinnabar in England, Holland, and different parts of Europe. The artificial, which was antiently called *minium*, a term now confined to red lead, does not differ from the natural in any quality essential to its value as a pigment; it varies in tint from dark red to scarlet; and both sorts are perfectly durable and unexceptionable pigments,—the most so perhaps of any we possess, when pure. It is true, nevertheless, that vermilions have obtained the double disrepute of fading in a strong light and of becoming black or dark by time and impure air: but colours, like characters, suffer contamination and disrepute from bad association; it has happened accordingly that vermilion which has been rendered lakey or crimson by mixture with lake or carmine, has faded in the light, and that when it has been toned to the scarlet hue by red or orange lead it has afterwards become blackened in impure air, &c. both of which adulterations were formerly practised, and hence the ill fame of vermilion both with authors and artists. We therefore repeat, that neither light, time, nor foul air, effect sensible change in true vermilions, and that they may be used safely in either water, oil, or fresco,—being colours of great chemical permanence, unaffected by other pigments, and among the least soluble of chemical substances.

Good vermilion is a powerful vivid colour of great body, weight, and opacity;—when pure it will be entirely decomposed and dissipated by fire in a red heat, and is therefore, in respect to the above mixtures, easily tested.

The following brilliant pigment from Iodine has been improperly called vermilion, and, if it should be used to dress or give unnatural vividness to true vermilion, may again bring it into disrepute. When red or orange lead has been substituted for or used in adulterating vermilion, muriatic acid applied to such pigments will turn them more or less white or gray; but pure vermilions will not be affected by the acid, nor will they by pure or caustic alkalis, which change the colour of the reds of Iodine. By burning more or less, vermilion may be brought to the colour of most of the red ochres.

II. IODINE SCARLET is a new pigment of a most vivid and beautiful scarlet colour, exceeding the brilliancy of vermilion. It has received several false appellations, but is truly an *Iodide* or *Bi-iodide of mercury*, varying in degrees of intense redness. It has the body and opacity of vermilion, but should be used with an ivory palette-knife, as iron and most metals change it to colours varying from yellow to black. Strong light rather deepens and cools it, and impure air soon utterly destroys its scarlet colour, and even metallizes it in substance. The charms of beauty and novelty have recommended it, particularly to amateurs; and its dazzling brilliancy might render it valuable for high and fiery effects of colour, if any mode of securing it from change should be devised; but it is out of the general scale of nature,— at any rate it should be used pure or alone, but in the recency of experience it ought not to be trusted by the artist. A similar pigment of a more crimson hue has lately appeared, having all the imperfections of the above, but in an inferior degree, and have both been improperly called vermilion, being among the most changeable of chemical substances. By time alone these colours vanish in a thin wash or glaze without apparent cause, and they attack almost every metallic substance, and some of them even in a dry state.

III. CHROMATE OF MERCURY. See Orange.

IV. RED LEAD, *Minium*, or *Saturnine Red*, by some old writers confounded with cinnabar, and called *Sinoper* or *Synoper*, is an oxide of lead of a scarlet colour and fine hue, warmer than common vermilion; bright, but not so vivid as the iodide of mercury, though it has the body and opacity of both these pigments, and has been confounded, even in name, with vermilion, with which it was formerly customary to mix it. When pure and alone, light does not affect its colour; but white lead, or any oxide or preparation of that metal mixed with it, soon deprives it of colour, as acids do also; and impure air will blacken and ultimately metallize it.

On account of its extreme fugitiveness when mixed with white lead, it cannot be used in tints; but employed, unmixed with other pigments, in simple varnish or oil not rendered drying by any metallic oxide, it may, under favourable circumstances, stand a long time; hence red lead has had a variable character for durability. It is in itself, however, an excellent dryer in oil, and has in this view been employed with other pigments; but, as regards colour, it cannot be mixed safely with any other pigments than the ochres, earths, and blacks in general: when employed in water the reds of lead, iodine, and mercury are warmed and brightened in colour by the addition of gum; this it does mechanically by allowing the darker particles to subside.

V. RED OCHRE is a name proper rather to a class, than to an individual pigment, and comprehends *Indian red*, *light red*, *Venetian red*, *scarlet ochre*, *Indian ochre*, *redding*, *ruddle*, *bole*, &c. beside other absurd appellations, such as *English vermilion* and *Spanish brown*, or *majolica*.

Almagra, the *Sil Atticum* of the antients, is a deep red ochre found in Andalusia, as is also their *terra Sinopica*, &c. or *Armenian bole*, dug originally in Cappadocia, and now found in New Jersey and elsewhere under the name of *blood-stone*.

The red ochres are, for the most part, rather hues and tints than definite colours, or more properly classed with the tertiary, semi-neutral, and broken colours; they are, nevertheless, often very valuable pigments for their tints in dead colouring, and for their permanence, &c. in water, oil, crayons, and fresco. The greater part of them are native pigments, found in most countries, and very abundantly and fine in our own; but some are productions of manufacture, and we have produced them in the variety of nature by art. The following are the most important of these pigments, most of which are available in enamel-painting.

1. INDIAN RED, according to its name, is brought from Bengal, and is a very rich iron ore, or per-oxide of iron. It is an anomalous red, of a purple-russet hue, of a good body, and valued when fine for the pureness and lakey tone of its tints. In a crude state it is a coarse powder, full of extremely hard and brilliant particles of a dark appearance, sometimes magnetic, and is greatly improved by grinding and washing over. Its chemical tendency is to deepen, nevertheless it is very permanent; neither light, impure air, mixture with other pigments, time, nor fire, effecting in general any sensible change in it; but, being opaque, and not keeping its place well, it is of course not fit for glazing. This pigment varies considerably in its hues,—that which is most rosy being esteemed the best, and affording the purest tints: inferior red ochres have been formerly substituted for it, and procured it a variable character, but it is now obtained abundantly, and may be had pure of respectable colourmen.

Persian Red is another name for this pigment, for which we have often heard the late Presidents, Mr. West, Sir Thomas Lawrence, and others, well experienced in its use, express the highest esteem.

2. LIGHT RED is an ochre of a russet-orange hue, principally valued for its tints. The common light red is brown ochre burnt, but the principal yellow ochres afford this colour best; and the brighter and better the yellow ochre is from which this pigment is prepared, the brighter will this red be, and the better flesh tints will it afford with white. There are, however, native ochres brought from India and other countries which supply its place, some of which are darkened by time and impure air; but in other respects light red has the general good properties of other ochres, dries admirably, and is much used both in figure and landscape. It affords also an excellent crayon.—See Orange Ochre.

3. TERRA PUZOLLI is a species of light red, as is also the

4. CARNAGIONE of the Italians, which differs from the above only in its hue, in which respect other variations and denominations are produced by dressing and compounding.

5. VENETIAN RED, or *Scarlet Ochre*. True Venetian red is said to be a native ochre, but the colours sold under this name are prepared artificially from sulphate of iron, or its residuum in the manufacturing of acids. They

are all of redder and deeper hues than light red, are very permanent, and have all the properties of good ochres.

Prussian red and *English red* are other names for the same pigment.

6. SPANISH RED is an ochre differing little from Venetian red.

VI. DRAGON'S BLOOD is a resinous substance, brought principally from the East Indies. It is of a warm semi-transparent, rather dull, red colour, which is deepened by impure air, and darkened by light. There are two or three sorts, but that in drops is the best. White lead soon destroys it, and it dries with extreme difficulty in oil. It is sometimes used to colour varnishes and lackers; but, notwithstanding it has been recommended as a pigment, it does not merit the attention of the artist.

VII. LAKE, a name derived from the *lac* or *lacca* of India, is the cognomen of a variety of transparent red and other coloured pigments of great beauty, prepared for the most part by precipitating coloured tinctures of dyeing drugs upon alumine and other earths, &c. The lakes are hence a numerous class of pigments, both with respect to the variety of their appellations and the substances from which they are prepared. The colouring matter of common lake is Brazil wood, which affords a very fugitive colour. Superior red lakes are prepared from cochineal, lac, and kermes; but the best of all are those prepared from the root of the *rubia tinctoria*, or madder plant. Of the various red lakes the following are the principal:—

1. RUBRIC, or MADDER LAKES. These pigments are of various colours, of which we shall speak at present of the red or rose colours only; which have obtained, from their material, their hues, or their inventor, the various names of rose rubiates, rose madder, pink madder, and Field's lakes.

The pigments formerly called madder lakes were brick-reds of dull ochrous hues; but for many years past these lakes have been prepared perfectly transparent, and literally as beautiful and pure in colour as the rose;[*] qualities in which they are unrivalled by the lakes and carmine of cochineal. The rose colours of madder have justly been considered as supplying a desideratum, and as the most valuable acquisition of the palette

---

[*] Of this we have ample evidence in the exquisite flowers of Hewlett, Bartholomew, Sintzenich, and others.

in modern times, since perfectly permanent transparent reds and rose-colours were previously unknown to the art of painting.

These pigments are of hues warm or cool, from pure pink to the deepest rose colour;—they afford the purest and truest carnation colours known;—form permanent tints with white lead; and their transparency renders them perfect glazing or finishing colours. They are not liable to change by the action of either light or impure air, nor by mixture with other pigments; but when not thoroughly edulcorated they are, in common with all lakes, tardy dryers in oil, the best remedy for which is the addition of a small portion of japanner's gold-size. Notwithstanding they are equally beautiful and durable as water-colours, they do not work therein with the entire fulness and facility of cochineal lakes : when, therefore, permanence is of no consideration, the latter may still be preferred; but in those works in which the hues and tints of nature are to be imitated with pure effect and permanence, the rose colours of madder are become indispensable, and their powers in these respects have been established by experience from the palettes of our first masters during upwards of a quarter of a century. With respect to the future too, there is this advantage attending these pigments, that they have naturally the peculiar quality of ultramarine, of improving in hue by time—their tendency being to their own specific prismatic red colour. These pigments have been imitated on the Continent with various success, and in many instances by the lakes of lac, cochineal, and carthamus. The best we have seen is the *laque de garance* of the French, the brightest of which was evidently tinged by the rouge of safflower, and proved inferior in durability to the genuine lake of madder. As, however, the colours of safflower, cochineal, and lac, are soluble in liquid ammonia and alkalis in general, which the true madder lakes or rubiates are not, the latter may be as easily tested by an alkali as ultramarine is by an acid; and if pure ammonia do not extract colour from a lake so tested, we may with general certainty pronounce it to be a true madder lake. See Madder Carmine 9.

2. LIQUID RUBIATE, or *Liquid Madder Lake*, is a concentrated tincture of madder of the most beautiful and perfect rose colour and transparency. It is used as a water-colour only in its simple state diluted with pure water, without gum; nevertheless, as it were by a sort of seeming caprice, it dries quicker in oil than alone or in water, by acting as a dryer to the oil. Mixed or ground with all other madder colours without gum, they form combinations which work freely in simple water, and produce the

most beautiful and permanent effects. The red of the definitive scale, Pl. 1. fig. 3. is of the pigments 1. and 2. combined. Liquid rubiate affords also a fine red ink,—and is a durable stain which bears washing, for marking, painting, or printing on cotton or linen cloth, &c.

3. SCARLET LAKE is prepared in form of drops from cochineal, and is of a beautiful transparent red colour, and excellent body, working well both in water and oil, though, like other lakes, it dries slowly. Strong light discolours and destroys it both in water and oil; and its tints with white lead, and its combinations with other pigments, are not permanent; yet when well prepared and judiciously used in sufficient body, and kept from strong light, it has been known to last many years; but it ought never to be employed in glazing, nor at all in performances that aim at high reputation and durability. It is commonly tinted with vermilion, which has probably been mixed with lakes at all times to give them scarlet hue and add to their weight; for upon examining with a powerful lens some fine pictures of antient masters, in which lake had been used in glazing, particles of vermilion were apparent, from which lake had evidently flown: unfortunately, however, these lakes are injured by vermilion as they are by lead, so that glazings of cochineal lake over vermilion or lead are particularly apt to vanish. This effect is very remarkable in several pictures of Cuyp, in which he has introduced a figure in red from which the shadows have disappeared, owing to their having been formed with lake over vermilion.

4. FLORENTINE LAKE differs from the last only in the mode of preparation, the lake so called having been formerly extracted from the shreds of scarlet cloth. The same may be said also of *Chinese Lake*.

5. HAMBURGH LAKE is a lake of great power and depth of colour, purplish, or inclining to crimson, which dries with extreme difficulty, but differs in no other essential quality from other cochineal lakes—an observation which applies to various lakes under the names of *Roman Lake*, *Venetian Lake*, and many others; not one of which, however respectively beautiful or reputed, is entitled to the character of durability either in hue, shade, or tint.

6. KERMES LAKE is the name of an antient pigment, perhaps the earliest of the European lakes; which name is sometimes spelt *cermes*, whence probably *cermosin* and *crimson*, and *kermine* or *carmine*. In some old books it is called *vermilion*, in allusion to the insect, or *vermes*, from which it is prepared; a title usurped probably by the sulphuret of mercury

or cinnabar, which now bears the name of vermilion. This lake is prepared from the kermes, which formerly supplied the place of cochineal. We have obtained the kermes from Poland, where it is still collected ; and some with which we have been favoured was brought from Cefalonia by Colonel C. J. Napier, who states that it is employed by the modern Greeks under the appellation κύπρο κόκινο, for dying their caps red. This substance and the lac of India probably afforded the lakes of the Venetian painters, and of the earliest painters in oil of the school of Van Eyck. Some old specimens of the pigment which we obtained were in drops of a powdery texture and crimson colour, warmer than cochineal lakes, and having less body and brilliancy, but worked well, and withstood the power of light better than the latter, though the sun ultimately discoloured and destroyed them. In all other respects they resemble the lakes of cochineal.

7. LAC LAKE is prepared from the lac or lacca of India, and is perhaps the first of the family of lakes. Its colouring matter resembles those of the cochineal and kermes, in being the production of a species of insects. Its colour is rich, transparent, and deep,—less brilliant and more durable than those of cochineal and kermes, but inferior in both these respects to the colours of madder. Used in body or strong glazing, as a shadow colour, it is of great power and much permanence; but in thin glazing it changes and flies, as it does also in tint with white lead.

A great variety of lakes, equally beautiful as those of cochineal, have been prepared from this substance in a recent state in India and China, many of which we have tried, and found uniformly less durable in proportion as they were more beautiful. In the properties of drying, &c. they resemble other lakes.

This appears to have been the lake which has stood best in old pictures, and was probably used by the Venetians, who had the trade of India when painting flourished at Venice. It is sometimes called *Indian Lake*.

8. CARMINE, a name originally given only to the fine feculences of the tinctures of kermes and cochineal, denotes generally at present any pigment which resembles them in beauty, richness of colour, and fineness of texture: hence we hear of blue and other coloured carmines, though the term is principally confined to the crimson and scarlet colours produced from cochineal by the agency of tin. These carmines are the brightest and most beautiful colours prepared from cochineal,—of a fine powdery texture and velvety richness. They vary from a rose colour to a warm red ; work admirably ;

and are in other respects, except the most essential—the want of durability, excellent pigments in water and oil:—they have not, however, any permanence in tint with white lead, and in glazing are soon discoloured and destroyed by the action of light, but are little affected by impure air, and are in other respects like the lakes of cochineal; all the pigments prepared from which may be tested by their solubility in liquid ammonia, which purples lakes prepared from the woods, but does not dissolve their colours.

9. MADDER CARMINE, or *Field's Carmine*, is, as its name expresses, prepared from madder. It differs from the rose lakes of madder principally in texture, and in the greater richness, depth, and transparency of its colour, which is of various hues from rose colour to crimson. These in other respects resemble the rubric or madder lakes, and are the only *durable carmines* for painting either in water or oil; for both which their texture qualifies them without previous grinding or preparation.—See Madder Lake, VII. 1.

10. ROSE PINK is a coarse kind of lake, produced by dying chalk or whitening with decoction of Brazil wood, &c. It is a pigment much used by paper-stainers, and in the commonest distemper painting, &c., but is too perishable to merit the attention of the artist.

11. ROUGE. The *Rouge Vegetale* of the French is a species of carmine, prepared from safflow or safflower, of exquisite beauty and great cost. Its principal uses consist in dyeing silks of rose colours, and in combining with levigated talc to form the paint of the toilette, or cosmetic colours employed by the fair, who would have

> Blooming youth and gay delight
> Sit on their rosy cheeks confess'd.

This pigment is, however, of too fugitive a colour to merit the attention of the artist, notwithstanding its great beauty, richness of colour, transparency, and free-working,—qualities which occasion it to be too often employed in heightening the apparent excellence of lakes and carmines. *Chinese Rouge* and *Pink Saucers* have much of these qualities, and appear to be prepared also from the carthamus or safflower.

VIII. RED ORPIMENT. See Orange Orpiment.

CHAP. XI.

OF BLUE.

*Where'er we gaze,—around, above, below,—
What rainbow tints, what magic charms are found!
Rocks, river, forest, mountains, all abound,
And bluest skies that harmonize the whole.*
BYRON'S CHILDE HAROLD.

THE third and last of the primary, or simple colours, is *blue,* which bears the same relation to shade that yellow does to light; hence it is the most retiring and diffusive of all colours, except purple and black: and all colours have the power of throwing it back in painting, in greater or less degree, in proportion to the intimacy of their relations to light; first white, then yellow, orange, red, &c.

Blue alone possesses entirely the quality technically called *coldness* in colouring, and it communicates this property variously to all other colours with which it happens to be compounded. It is most powerful in a strong light, and appears to become neutral and pale in a declining light, owing to its ruling affinity with black or shade, and its power of absorbing light: hence the eye of the artist is liable to be deceived when painting with blue in too low a light, or toward the close of day, to the endangering of the warmth and harmony of his picture. Blue enters into combination with yellow in the composition of all *greens,* and with red in all *purples;* it characterizes the tertiary *olive,* and is also the prime colour or archeus of the neutral *black,* &c., and also of the semineutral *slate colour,* &c.: hence blue is changed in hue less than any other colour by mixture with black, as it is also by distance. It enters also subordinately into all other tertiary and broken colours, and, as nearest in the scale to black, it breaks and contrasts powerfully and agreeably with white, as in watchet or pale blues, the sky, &c. It is less active than the other primaries in reflecting light, and therefore is sooner lost as a local colour by assimilation with distance. It is an antient doctrine that the azure of the sky is a compound of light and darkness, and some have argued hence that blue is not a primary colour, but a compound of black and white; but pure or *neutral* black and white compound in infinite shades, all of which are neutral also, or *grey*. It is true that a mixture of black and white is of a *cool* hue, because black is not a primary colour but a compound of the three primary colours in which blue predominates, and this predominance is rendered more sensible when black is diluted with white. As to the colour of the sky, in which light and shade are compounded, it is neutral also, and never blue except by contrast: thus, the more the light of the sun partakes of the golden or orange hue, and the more parched and burnt the earth is, the bluer appears the sky, as in Italy, and all hot countries. In England, where the sun is cooler, and a perpetual verdure reigns, infusing blue latently into the landscape, the sky is warmer and nearer to neutrality, and partakes of a diversity of *grays*, beautifully melodizing with blue as their key, and harmonizing with the light and the landscape. Thus the colour of the sky is always a contrast to the direct and reflected light of the scene: if therefore this light were of a *rose-colour*, the neutral of the sky would be converted into green; or if the light were *purple*, the sky would become *yellow*, and such would it be in all other cases, according to the laws of chromatic equivalence and contrast, as often appears in the openings of coloured clouds at the rising and setting of the sun.

Blue is discordant in juxtaposition with green, and in a less degree so with purple, both which are cool colours, and therefore blue requires its contrast, *orange*, in equal proportion, either of surface or intensity, to compensate or resolve its dissonances and correct its coldness: of all colours, except black, it contrasts white most powerfully. In all harmonious combinations of colours, whether of mixture or neighbourhood, blue is the natural, prime, or predominating power: accordingly blue is in colouring what the note C is in music,—the natural key, archeus or ruling tone, universally agreeable to the eye, when in due relation to the composition, and may be more frequently repeated therein, purely or unbroken, than either of the other primaries: this is however a matter of taste, as in music, and subject to artificial rules founded on the laws of chromatic combination.

The moral expression, or effects of blue, or its influence on the feelings and passions, partake of its cold and shadowy relations in soothing and inclining to melancholy, and its allied sentiments: accordingly it is rather a sedate than a gay colour, even when in its utmost brilliancy. In nature it is the colour of Heaven and of the eye, and thence emblematical of intelligence and divinity. It is accordingly, by a natural analogy, used in mythological representations to distinguish the mantle of Minerva, the blue-eyed goddess, and the veil of Juno, the goddess of air; while Diana, or the Moon, is robed in blue and white, as the Isis of the Egyptians and her priests were in pure azure; and Poetry herself is personified in a vesture of celestial blue.

These analogies and effects of colours are by no means to be disregarded, since they are as various, simply and conjointly, as are colours and their tints; and it is by attention to them that colour conduces to sentiment and expression in painting. Even where the symbolical uses of colours are merely fanciful or conventional, they are not to be totally rejected, since, by association and common consent, they acquire arbitrary signification in the manner of words: it is their natural expression, however, which principally claims our attention. Indeed a just knowledge of the relations of colours, and their effects upon the passions, feelings, and intellect, seems, we repeat, hardly less essential to the poet than to the painter: hence he often employs their ideas and terms with happy effect, and as frequently fails when wanting a proper comprehension, feeling, or taste of their powers: the latter case is however more fatal to the artist than to the poet. The following are instances in illustration of these effects, and of the coincidences of poetry and painting in the use of the colour blue.

As soothing, sedate, or sad,—

> Long, *Pity*, let the nations view
> Thy sky-worn robes of *tend'rest blue*.
> COLLINS.

> Ah! hills beloved! where once a happy child,
> Your beechen *shades*, your turf, your flow'rs among
> I wove your *blue-bells* into garlands wild,
> And 'woke your echoes with my artless song.
> CHARLOTTE SMITH.

> And heal the harms of thwarting thunder *blue*.
> MILTON.

As intellectual, &c.

> The *blue-eyed* progeny of Jove.
> DRYDEN.

> The *joyful sun* sprung up the *blue serene*.
> AKENSIDE.

In accordance with *white*, &c.

> *White* and *azure*, laced
> With *blue* of heaven's own tinct.
> SHAKSPEARE'S CYMBELINE, Act II. Sc. 2.

> I saw their thousand years of *snow*
> On high—their wide long lake below,
> And the *blue* Rhone in fullest flow.
> BYRON'S PRISONER OF CHILLON.

> Morning *light*
> More *orient* in yon western cloud that draws
> O'er the *blue* firmament a radiant *white*.
> MILTON.

> As the *pearl*
> Shines in the concave of its *azure* bed.
> AKENSIDE'S PLEASURES OF IMAGINATION, line 454.

> In *azure* channels glide on *silver* sands.
> FLETCHER.

> The *daisy, primrose, violet darkly blue*,
> And *polyanthus of unnumber'd dyes*.
> THOMSON.

The poets almost invariably use blue with its proper contrast; thus—

> But Fame, with *golden wings*, aloft doth fly
>   Above the reach of ruinous decay,
> And with brave plumes doth beat the *azure sky*,
>   Admired of base-born men from far away.
> SPENSER'S RUINS OF TIME.

> As then no wind at all there blew,
>   No swelling cloud accloid the air,
> The *sky, like glass of watchet hue*,
>   Reflected Phœby's *golden haire;*
> The garnisht trees no pendant stirr'd,
> No voice was heard of any bird.
> IDEM, ELEGY ON SIR PHILIP SIDNEY.

> His *golden locks* waved wild and free
> O'er his *blue eyes*.
> MRS. PICKERSGILL.

> Their eyes *blue languish*, and their *golden hair*—.
> COLLINS.

> There's *gold*, and here my *bluest* veins to kiss.
> SHAKSPEARE, ANTONY AND CLEOPATRA.

> Hast thou left thy *blue* course in heaven, *golden-haired* son of the sky?
> OSSIAN.

In the following, white is blended in the contrast, &c.

> Rise then, *fair blue-eyed* maid, rise and discover
> Thy *silver brow*, and meet thy *golden lover*.
> CRASHAW.

> Why does one climate and one soil endue
> The *blushing poppy* with an *orange hue*,
> Yet leave the *lily pale*, and tinge the *violet blue?*
> PRIOR.

And in the succeeding examples it is associated with black or shade, melancholy and coldness, &c.

> O coward Conscience! how dost thou afflict me!
> The lights burn *blue?*—It is now *dead midnight.*—
> *Cold fearful* drops stand on my trembling flesh.
> SHAKSPEARE, RICHARD THE THIRD.

> Common mother, thou,
> Whose womb unmeasurable, and infinite breast,
> Teems and feeds all; whose self-same mettle,
> Whereof thy proud child, arrogant man, is puff'd,
> Engenders the *black* toad and adder *blue*,
> The *gilded* newt and eyeless venom'd worm.
> IDEM, TIMON, Act IV. Scene 3.

The paucity of blue pigments, in comparison with those of yellow and red, is amply compensated by their value and perfection; nor is the palette deficient in pigments of this colour, of which the following comprise all that are in any respect of importance to the painter.

I. ULTRAMARINE, *Lazuline, Azurine,* or *Azure,* is prepared from the lapis lazuli, a precious stone found principally in Persia and Siberia. It is the most celebrated of all modern pigments, and, from its name and attributes, is probably the same as the no less celebrated *Armenian blue*, or *Cyanus*, of the antients. Of the latter, Theophrastus informs us that the honour of inventing its factitious preparation (by perhaps the very singular chemico-mechanical process still in use for ultramarine) was ascribed in the Egyptian annals to one of their kings;* and it was so highly prized that the Phœnicians paid their tribute in it, and it was given in presents to princes: hence it was a common practice in those times to counterfeit it. Our opinion of the identity of these pigments is considerably strengthened by the accounts modern travellers give of the brilliant blue painting still remaining in the ruins of temples in Upper Egypt, which is described as having all the appearance of ultramarine. Add to this, also, that the Chinese have the art of preparing this pigment; and as they are imitators, and rarely inventors, and cannot be supposed to have learnt it of the Europeans, it is to be inferred that they possess it as an antient art: that they have it, we conclude from having received specimens of this pigment, of a good colour, direct from Canton. In China, too, the lapis lazuli is highly esteemed, and is worn by mandarines as badges of nobility conferred only by the emperor; which remarkably coincides with the antient usage related by Theophrastus.

Ultramarine has not obtained its reputation upon slight pretensions, being, when skilfully prepared, of the most exquisitely beautiful blue, varying from the utmost depth of shadow to the highest brilliancy of light and colour,—transparent in all its shades, and pure in its tints. It is of a true medial blue, when perfect, partaking neither of purple on the one hand, nor of green on the other: it is neither subject to injury by damp and impure air, nor by the intensest action of light, and it is so eminently permanent that it remains perfectly unchanged in the oldest paintings; and there can be little doubt that it is the same pigment which still continues with all its original force and beauty in the temples of Upper Egypt, after an exposure of at least three thousand years. The antient Egyptians had, however, other blues, of which we have already mentioned their counterfeit Armenian blue; and we have lately seen some balls of blue pigment, of considerable depth and purity of colour, in the collection of Mr. Sams, obtained by him from the ruins of Upper Egypt, which is probably of the

---

* Theophrast. de Lapid. xcviii. Plin. lib. xxxvii.

same kind. The Egyptians had also several vitreous blues, with which they decorated their figures and mummies.

Ultramarine dries well, works well in oil and fresco, and neither gives nor receives injury from other good pigments. It has so much of the quality of light in it, and of the tint of air,—is so purely a sky-colour, and is hence so singularly adapted to the direct and reflex light of the sky, and to become the antagonist of sunshine,—that it is indispensable to the landscape-painter; and it is so pure, so true, and so unchangeable in its tints and glazings, as to be no less essential in imitating the exquisite colouring of nature in flesh and flowers.

To this may be added, that it enters so admirably into purples, blacks, greens, grays, and broken colours, that it has justly obtained the reputation of clearing or carrying light and air into all colours, both in mixture and glazing, and a sort of claim to universality throughout a picture. These qualities of ultramarine are admirably illustrated by an experiment recorded in our next article.

This is the sober character of perfect ultramarine, and no eulogy, any farther than truth when attendant on merit, is ever the most powerful of all praise.

It is true, nevertheless, that ultramarine is not always entitled to the whole of this commendation, and it is necessary that the artist should be thoroughly and in every respect acquainted with a pigment of such importance; we will therefore take a view of the imperfections to which it is liable. Ultramarine is often *coarse in texture;* and to this there are temptations in the making, because in this way it is more easy of preparation, and more abundant, deep, and valuable *in appearance:* yet such ultramarine cannot be used with effect, nor ground fine without injuring its colour. Again, it is rarely separated in a pure state from the lapis lazuli, which is an exceedingly varying and compound mineral, abounding with earthy and metallic parts in different states of oxidation and composition: hence ultramarine sometimes contains iron in the state of red oxide, and such ultramarine has a *purple* cast; and sometimes it contains the same metal in the state of yellow oxide, and then it is of a *green* tone: it has still more frequently in it a portion of the sulphuret of iron, which is black, and gives to the ultramarine a deeper but a *dusky hue.* Some artists, nevertheless, have preferred ultramarine for each of these tones; yet are they imperfections which may account for various effects and defects of this pigment in painting. *Grow-*

ing deeper by age has been attributed to ultramarine, but it is such specimens alone as would acquire depth in the fire that could be subject to such change; and it has been justly supposed that in pictures wherein other colours have failed by age, it may have taken this appearance by contrast. Ultramarine prepared from calcined lapis lazuli is not subject to deepen by age; but this advantage is purchased by some sacrifice of the vivid, warm, and pure azure colour of ultramarine prepared from the unburnt stone: the perfection of the pigment is also in a great measure dependent upon the quality of the lapis lazuli from which it is prepared. As a precious material, ultramarine has been subjected to *adulteration,* and it has been dyed, damped, and oiled, to enrich its appearance; but these attempts of fraud may be easily detected, and the genuine may as easily be distinguished from the spurious by dropping a few particles of the pigment into lemon-juice, or any other acid, which almost instantly destroys the colour of the true ultramarine totally,—an example of the subtile power of chemistry in changing instantly one of the most permanent substances in nature, and a reason why the mineral itself is so rare among her productions.

Though unexceptionable as an oil-colour, both in solid painting and glazing, it does not work so well as some other blues in water; but when extremely fine in texture, or when a considerable portion of gum, which renders it transparent, can be used with it to give it connexion or adhesion while flowing, it becomes a pigment no less valuable in water painting than in oil; but little gum can however be employed with it when its vivid azure is to be preserved, as in illuminated manuscripts and missals. The blue of the Definitive Scale, Pl. 1. fig. 3. is of the middle depth of ultramarine.

Such are the principal merits and defects of ultramarine as a blue colour; and the fine greens, purples, and grays of the old masters, are often unquestionably compounds of ultramarine; and it was the only blue formerly used in fresco.

The immense price of ultramarine in former times * was almost a prohibition to its use; but the spirit of modern commerce having supplied its

---

* Walpole relates that "Charles I. presented to Mrs. Carlisle five hundred pounds' worth of ultramarine, which lay in so small a compass as only to cover his hand."—ANECDOTES OF PAINTING. Barrow states, however, that Vandyke partook of the King's present equally with Mrs. Carlisle.—DICT. POLYGRAPH.

material more abundantly, and the discoveries and improvement of Prussian, Antwerp, and cobalt blues, having furnished substitutes for its ordinary uses, it may now be obtained at moderate prices, particularly its lighter and very useful tints.

Pure ultramarine varies in shade from light to dark, and in hue from pale warm azure to the deepest cold blue; the former of which, when impure in colour, is called *ultramarine ashes*.

II. FACTITIOUS ULTRAMARINE, *French Ultramarine, Outremère de Guimet, Bleu de Garance*, &c. In some of the latter numbers of " Brande's Journal," are accounts of a process for producing *factitious* ultramarine; and a variety of these have been before the public under the above and other names. These pigments are in general of deep rich blue colours, darker and less azure than fine ultramarine of the same depths, and answering to the same acid tests, but variously affected by fire and other agents: none of them, however, possess the merits of genuine ultramarine, and their relative value to other blues remains to be determined by mature experience. An experiment in which this blue was tried by an ingenious artist and friend of the author, however, speaks little in its favour. He took a picture, the sky of which had been recently painted in the ordinary manner with *Prussian blue* and white; and having painted on the clear part of the sky uniform portions with tints formed of the best *factitious ultramarine, cobalt blue*, and *genuine ultramarine*, so as to match the ground of the sky, and to disappear to the eye thereon by blending with the ground, when viewed at a moderate distance, he set the picture aside for some months; after which it appeared upon examination that the colour of these various blue pigments had taken different ways, and departed from the hue of the ground:—the factitious ultramarine had *blackened*,—the cobalt blue *greened*,—the *true ultramarine* appeared of a *pure azure*, like a spot of light,—and their ground, the Prussian blue sky, appeared by contrast with the ultramarine of a *gray* or *slate colour*.

All the chemical combinations of iron with inflammable bases, under proper management, afford blue colours; and in this respect the factitious ultramarines and Prussian blue are analogous.

III. 1. COBALT BLUE is the name now appropriated to the modern improved blue prepared with metallic cobalt, or its oxides, although it properly belongs to a class of pigments including *Saxon blue, Dutch ultramarine, Royal blue, Hungary blue, Smalt, Zaffre* or *Enamel blue,* and *Dumont's blue.* These differ principally in their degrees of purity, and the nature of the earths with which they are compounded.

The first is the finest cobalt blue, and may not improperly be called a blue lake, the colour of which is brought up by fire, in the manner of enamel blues; and it is, when well prepared, of a pure blue colour, neither tending to green nor purple, and approaching in brilliancy to the finest ultramarine. It has not however the body, transparency, and depth, nor the natural and modest hue, of the latter; yet it is superior in beauty to all other blue pigments. Cobalt blue works better in water than ultramarine in general does, and is hence an acquisition to those who have not the management of the latter, and also on account of its cheapness. It resists the action of strong light and acids; but its beauty declines by time, and impure air greens and ultimately blackens it.

It dries well in oil, does not injure nor suffer injury from pigments in general, and may be used with a proper flux in enamel painting, and perhaps also in fresco.

Various appellations have been given to this pigment from its preparers and venders, and it has been called *Vienna blue, Paris blue, azure,* and, very improperly, *ultramarine.*

2. SMALT, sometimes called *Azure,* is an impure vitreous cobalt blue, prepared upon a base of silex, and much used by the laundress for neutralizing the tawny or Isabella colour of linen, &c., under the name of *Powder-blue.* It is in general of a coarse gritty texture, light blue colour, and little body. It does not work so well as the preceding, but dries quickly, and resembles it in other respects;—it varies, however, exceedingly in its qualities; and the finer sorts, called *Dumont's Blue,* which is employed in water-colour painting, is remarkably rich and beautiful.

3. ROYAL BLUE is a deeper coloured and very beautiful smalt, and is also a vitreous pigment, principally used in painting on glass and enamel, in which uses it is very permanent; but in water and oil its beauty soon decays, as is no uncommon case with other vitrified pigments; and it is not in other respects an eligible pigment, being, notwithstanding its beautiful appearance, very inferior to other cobalt blues.

IV. 1. PRUSSIAN BLUE, otherwise called *Berlin blue, Parisian blue,*

Cyanide of Iron, &c., is rather a modern pigment, produced by the combination of the prussic or hydro-cyanic acid and iron. It is of a deep and powerful blue colour, of vast body and considerable transparency, and forms tints of much beauty with white lead, though they are by no means equal in purity and brilliancy to those of cobalt and ultramarine, nor have they the perfect durability of the latter.

Notwithstanding Prussian blue lasts a long time under favourable circumstances, its tints fade by the action of strong light, and it is purpled or darkened by damp or impure air. It becomes greenish also sometimes by a developement of the yellow oxide of iron. The colour of this pigment has also the singular property of fluctuating or of going and coming under some changes of circumstances, and time has a neutralizing tendency upon its colour.

It dries and glazes well in oil; but its great and principal use is in painting deep blues, in which its body secures its permanence, and its transparency gives force to its depth. It is also valuable in compounding deep purples with lake, and is a powerful neutralizer and component of black, and adds considerably to its intensity. It is a pigment much used in the common offices of painting, in preparing blues for the laundress, and in dyeing. Mineralogists speak of a *Native Prussian Blue.*

2. ANTWERP BLUE is a lighter-coloured and somewhat brighter Prussian blue, or ferro-prussiate of alumine, having more of the terrene basis, but all the other qualities of that pigment, except its extreme depth. *Haerlem Blue* is a similar pigment.

V. 1. INDIGO, or *Indian Blue,* is a pigment manufactured in the East and West Indies from several plants, but principally from the anil or indigofera. It is of various qualities, and has been long known, and of great use in dyeing. In painting it is not so bright as Prussian blue, but is extremely powerful and transparent; hence it may be substituted for some of the uses of Prussian blue. It is of great body, and glazes and works well both in water and oil. Its relative permanence as a dye has obtained it a false character of extreme durability in painting, a quality in which it is nevertheless very inferior even to Prussian blue.

It is injured by impure air, and in glazing some specimens are firmer than others, but not durable;—in tint with white lead they are all fugitive; when used, however, in considerable body in shadow, it is more permanent, but in all respects inferior to Prussian blue.

2. INTENSE BLUE is indigo refined by solution and precipitation, in which state it is equal in colour to Antwerp blue. By this process indigo also becomes more durable, and much more powerful, transparent, and deep. It washes and works admirably in water;—in other respects it has the common properties of indigo. We have been assured by an eminent architect, equally able and experienced in the use of colours, that these blues of indigo have the property of pushing or detaching Indian ink from paper. The same is supposed to belong to other blues; but as this effect is chemical, it can hardly be an attribute of mere colour.

VI. 1. BLUE VERDITER is a blue oxide of copper, or precipitate of the nitrate of copper by lime, and is of a beautiful light blue colour. It is little affected by light; but time, damp, and impure air turn it green, and ultimately blacken it,—changes which ensue even more rapidly in oil than in water: it is therefore by no means an eligible pigment in oil, and is principally confined to distemper painting and the uses of the paper-stainer, though it has been found to stand well many years in water-colour drawings and in crayon paintings, when preserved dry.

2. SAUNDERS BLUE, a corrupt name, from *Cendres Bleus*, the original denomination probably of *ultramarine ashes*, is of two kinds, the natural and the artificial: the artificial is a verditer prepared by an alkali from sulphate of copper; the natural is a blue mineral found near copper-mines, and is the same as—

3. MOUNTAIN BLUE, found in similar situations as the above. A very beautiful substance of this kind, a *carbonate of copper*, both blue and green, is found in Cumberland. None of these blues of copper are, however, durable: used in oil, they become green, and as pigments are precisely of the character of verditers.

4. SCHWEINFURT BLUE appears to be the same in substance as Scheele's green, prepared without heat or treated with an alkali. It is a beautiful colour, liable to the same changes, and is of the same habits as blue verditer and the above pigments.

VII. BLUE BICE, *Iris* or *Terre Bleu*, is sometimes confounded with the above copper blues; but the true bice is said to be prepared from the *lapis Armenius* of Germany and the Tyrol, and is a light bright blue. The true Armenian stone of the antients was probably the lapis lazuli of later times, and the blue prepared therefrom the same as our ultramarine.

Ground smalts, blue verditer, and other pigments, have passed under the name of bice, which has therefore become a very equivocal pigment, and its name nearly obsolete; nor is it at present to be found in the shops.

VIII. BLUE OCHRE is a mineral colour of rare occurrence, found in Cornwall, and also in North America, and is a *sub-phosphate of iron*. What Indian red is to the colour red, and Oxford ochre to yellow, this pigment is to the colour blue; they class in likeness of character;—hence it is admirable rather for the modesty and solidity than for the brilliancy of its colour. It has the body of other ochres, more transparency, and is of considerable depth. It works well both in water and oil, dries readily, and does not suffer in tint with white lead, nor change when exposed to the action of strong light, damp or impure air: it is therefore, as far as its powers extend, an eligible pigment, though it is not in general use, nor easily procurable. It answers to the same acid tests as ultramarine, and is distinguishable from it by changing from a blue phosphate to an olive-brown ochrous oxide of iron when exposed to a red heat. It has been improperly called *native Prussian blue*.

IX. BLUE CARMINE is a blue oxide of molybdena, of which little is known as a substance or as a pigment. It is said to be of a beautiful blue colour, and durable in a strong light, but is subject to be changed in hue by other substances, and blackened by foul air: we may conjecture, therefore, that it is not of much value in painting.

CHAP. XII.

OF THE SECONDARY COLOURS.

OF ORANGE.

> Bear me to the *citron* groves—
> To where the *lemon* and the piercing *lime*,
> With the *deep orange* glowing through the *green*,
> Their lighter glories blend.
>
> THOMSON.

ORANGE is the first of the secondary colours in relation to light, being in all the variety of its hues composed of *yellow* and *red*. A true or perfect orange is such a compound of red and yellow as will neutralize a perfect blue in equal quantity either of surface or intensity, and the proportions of such compound are five of perfect red to three of perfect yellow. When orange inclines to red, it takes the names of *scarlet, poppy, coquilicot*, &c. In *gold* colour, &c. it leans toward yellow. It enters into combination with green in forming the tertiary *citrine*, and with purple it constitutes the tertiary *russet:*—it forms also a series of warm semineutral colours with *black*, and harmonizes in contact and variety of tints with *white*.

Orange is an advancing colour in painting :—in nature it is effective at a great distance, acting powerfully on the eye,—diminishing its sensibility in proportion to the strength of the light in which it is viewed; and it is of the hue and partakes of the vividness of sunshine, as it does also of all the powers of its components, red and yellow.

This secondary is pre-eminently a *warm* colour, being the equal contrast or antagonist in this respect, as it is also in colour, to blue, to which the attribute of *coolness* peculiarly belongs:—hence it is discordant when standing alone with yellow or with red, unresolved by their proper contrasts, or harmonizing colours, purple and green.

OF THE SECONDARY COLOURS.—OF ORANGE.

As an archeus, or ruling colour, orange corresponds to the key of F in music, and it is one of the most agreeable keys or archeï in toning a picture, from the richness and warmth of its effect;—accordingly its influence on feeling and the mind is gay and cheerful, and opposed to the soothing and sedate.

In the well-known fruit of the aurantium, called *orange* from its golden hue, from which fruit this colour borrows its well-adapted name, nature has associated two primary colours with two primary tastes which seem to be analogous,—a red and yellow compound colour, with a sweet and acid compound flavour.

The poets confound orange with its ruling colour yellow, and, by a metonymy, use in its place the terms golden, gilding, orient, &c., to express the signification of this colour in constructing of tropes; and it appears to be hardly less effective and necessary to warmth of description with the poet than with the painter, of which our preceding quotations afford instances; and some of the following illustrations of the poetic employment of this colour are in point. As according with light, &c.—

> So sweet a kiss the *golden sun* gives not
> To those *fresh morning drops upon the rose*,
> Nor shines the *silver moon* one half so *bright*
> Through the *transparent* bosom of the deep.
> SHAKSPEARE.

> No more the *rising sun* shall *gild the morn*,
> Nor *evening Cynthia* fill her *silver* horn.
> POPE.

> Reclining soft on many a *golden cloud*.
> ROWE.

> Heaven's *golden-wing'd* herald.
> CRASHAW.

> *Orient* liquors in a *crystal* glass.
> MILTON.

> Extremes, alike, in either hue behold,
> Hot—in the *golden*, in the *silvery—cold*.
> SHEE'S ELEMENTS, Canto v. l. 310.

As harmonizing with its co-secondaries:—

> Culls the delicious fruit that hangs in air,
> The *purple* plum, *green* fig, or *golden* pear.
> ROGERS.

As contrasted with blue:—

> From *golden cups* or hare-bells *blue*.
> MRS. PICKERSGILL.

Of this relation we have given a number of examples in the preceding chapter;—in the following description, by Sir Humphrey Davy, we have an instance from nature of the various relations of this colour associated:— From the summit of Vesuvius " we see the rich fields covered with flax, maize, or millet, and intersected by rows of trees, which support the *green* and graceful festoons of the vine; the *orange and lemon trees covered with golden fruit* appear in the sheltered glens; the olive trees cover the lower hills; islands, *purple in the beams of the setting sun*, are scattered over the sea in the west; and the sky is *tinted with red, softening into the brightest and purest azure*."—LAST DAYS OF A PHILOSOPHER.

And Shakspeare thus employs this colour in accordance with black:—

> The ousel-cock, so *black of hue*,
> With *orange-tawny* bill.

Butler also uses the same compound epithet:—

> At that an egg, let fly,
> Hit him directly o'er the eye,
> And running down his cheek, besmear'd
> With *orange-tawny* slime his beard.
> HUDIBRAS, Part I. Canto II.

The list of original orange pigments is so deficient, that in some treatises on the subject of colours, orange is not even named as a colour. This may have arisen partly from the unsettled signification of the term; partly from improperly calling these pigments reds, yellows, &c.; and partly also from their orginal paucity. The following are accordingly all the pigments in general use which can properly be classed under the name of orange, though most of them are called reds or yellows:—

I. MIXED ORANGE. Orange being a colour compounded of red and yellow, the place of original orange pigments may be supplied by mixture of the two latter colours; by glazing one over the other; by stippling, or other modes of breaking and intermixing them in working, according to the nature of the work and the effect required. For reasons before given, mixed pigments are inferior to the simple or homogeneous in colour, working, and other properties: yet some pigments mix and combine more cordially and with better results than others; this is the case with the *liquid rubiate* and *gamboge*, and they form the best and most durable mixed orange of all hues for painting in water. In oil the compounding of colours is more easily effected.

II. ORANGE VERMILION is a sulphuret of quicksilver or vermilion of an orange colour, newly introduced: it resembles red lead in appearance, but is not subject to its changes, being a perfectly durable pigment under every circumstance of oil or water painting. Its tints are much warmer than those of red or orange lead; and it is a most powerful tinger of white, yielding purer and more delicate warm carnation tints than any known pigment, and much resembling those of Titian and Rubens. It is the best and only unexceptionable orange we possess, drying in simple linseed oil, and having the powerful body and properties of the other vermilions, and may be tested in the same manner. It works with best effect in water with a considerable portion of gum. The orange of the definitive scale, Plate i. fig. 3. is of this pigment.

III. 1. CHROME ORANGE is a beautiful orange pigment, and is one of the most durable and least exceptionable chromates of lead, and not of iron, as it is commonly called, being truly a subchromate of lead.

It is, when well prepared, of a brighter colour than red, or orange vermilion, but is inferior in durability and body to the latter pigment, being liable to the changes and affinities of the chrome yellows in a somewhat less degree, but less liable to change than the orange oxide of lead (v.) following.

2. LAQUE MINERAL is a French pigment, a species of chromic orange, similar to the above. This name is also given to orange oxide of iron.

3. CHROMATE OF MERCURY is improperly classed as a red with vermilion, for though it is of a bright ochrous red colour in powder, it is, when ground, of a bright orange ochre colour, and affords with white very pure orange-coloured tints. Nevertheless it is a bad pigment, since light soon changes it to a deep russet colour, and foul air reduces it to extreme blackness.

IV. 1. ORANGE OCHRE, called also *Spanish ochre*, &c. is a very bright yellow ochre burnt, by which operation it acquires warmth, colour, transparency, and depth. In colour it is moderately bright, dries and works well both in water and oil, and is a very durable and eligible pigment. It may be used in enamel-painting, and has all the properties of its original ochre in other respects. See Yellow Ochre.

2. JAUNE DE MARS is an artificial iron ochre, similar to the above, of which we formerly prepared a variety brighter, richer, and more transparent than the above.

3. DAMONICO, or *Monicon*, is also an iron ochre, being a compound of Terra di Sienna and Roman ochre burnt, and having all their qualities. It is rather more russet in hue than the above, has considerable transparency, is rich and durable in colour, and affords good flesh tints.

4. BURNT SIENNA EARTH is, as its name expresses, the *Terra di Sienna* burnt, and is of an orange russet colour. What has been said of orange ochre and Damonico may be repeated of burnt Sienna. It is richer in colour, deeper, and more transparent, and works better than *raw Sienna earth;* but, in other respects, has all the properties of its parent colour, and is permanent and eligible wherever it may be useful.

5. *Light Red* and *Venetian Red*, before treated of, are also to be considered as impure, but durable, orange colours ; and several artificial preparations of iron afford excellent colours of this class.

V. ORANGE LEAD is an oxide of lead of a more vivid and warmer colour than *red lead*, but, in other respects, does not differ essentially from that pigment.

VI. ORANGE ORPIMENT, or *Realgar*, improperly called also *Red Orpiment*, since it is of a brilliant orange colour, inclining to yellow. There are two kinds of this pigment; the one, *native*, the other, *factitious;* the first of which is called *sandarach*, &c., and is of rather a redder colour than the factitious. They are the same in qualities as pigments, and differ not otherwise than in colour from *yellow orpiment*, to which the old painters gave the orange hue by heat, and then called it *alchymy*.

VII. GOLDEN SULPHUR OF ANTIMONY is a *Hydro-sulphuret of*

antimony, of an orange colour, which is destroyed by the action of strong light. It is a bad dryer in oil, injurious to many colours, and in no respect an eligible pigment either in oil or water.

VIII. MADDER ORANGE, or *Orange Lake*, is a madder lake of an orange hue, varying from yellow to rose-colour and brown. This variety of madder colours differs not essentially in other respects from those of which we have already spoken, except in a tendency toward redness in the course of time.

IX. ANOTTA, *Arnotta, Annotto, Terra Orleana, Roucou*, &c. are names of a vegetal substance brought from the West Indies, of an orange-red colour, soluble in water and spirit of wine, but very fugitive and changeable, and not fit for painting. It is principally used by the dyer, and in colouring cheese. It is also an ingredient in some lacquers. See Carucru, Ch. xix. art. ii.

CHAP. XIII.

OF GREEN.

*But where fair Isis rolls her purer wave
The partial muse delighted loves to lave;
On her green banks a greener wreath is wove,
To crown the bards that haunt her classic grove.*
BYRON.

GREEN, which occupies the middle station in the natural scale of colours and in relation to light and shade, is the second of the secondary colours: it is composed of the extreme primaries, *yellow* and *blue*, and is most perfect in hue when constituted in the proportions of *three* of yellow to *eight* of blue of equal intensities; because such a green will perfectly neutralize and contrast a perfect red in the proportions of *eleven* to *five*, either of space or power, as adduced on our Scale of Chromatic Equivalents. Green, mixed with orange, converts it into the one extreme tertiary *citrine*; and, mixed with purple, it becomes the other extreme tertiary, *olive;* hence its relations and accordances are more general, and it contrasts more agreeably with all colours than any other individual colour. It has accordingly been adopted with perfect wisdom in nature as the general garb of the vegetal creation. It is indeed in every respect a central or medial colour, being the contrast, compensatory in the proportion of eleven to five, of the middle primary, *red*, on the one hand, and of the middle tertiary, *russet*, on the other; and, unlike the other secondaries, all its hues, whether tending to blue or yellow, are of the same denomination.

These attributes of green, which render it so universally effective in contrasting of colours, cause it also to become the least useful in compounding them, and the most apt to defile other colours in mixture: nevertheless it forms valuable semi-neutrals of the *olive* class with *black*, for of such subdued tones are the greens, by which the more vivid hues of nature are contrasted; accordingly the various greens of foliage are always more or less semi-neutral in colour. As *green* is the most general colour of vegetal nature, and principal in foliage; so *red*, its harmonizing colour, and compounds of red, are most general and principal in flowers. *Purple* flowers are commonly contrasted with centres or variegations of bright *yellow*, as *blue* flowers are with like relievings of *orange*; and there is a prevailing hue, or character, in the green colour of the foliage of almost every plant, by which it is harmonized with the colours of its flowers; so also—

> No tree in all the grove but has its charms,
> Though each its hue peculiar; *paler* some,
> And of a *warmish gray*; the willow such,
> And poplar, that with *silver* lines his leaf,
> And ash, far stretching his *umbrageous* arms;
> Of *deeper green* the elm; and *deeper still*,
> Lord of the woods, the long-surviving oak.
> ———— Not unnoticed pass
> The sycamore, capricious in attire;
> Now *green*, now *tawny*, and, ere autumn yet
> Have changed the woods, in *scarlet honours bright*.
>
> COWPER.

These changes of the leaf may be attributed, like those of flowers, to the various action of the *oxygenous principle* in light and air upon the carbon, or hydrogenous *principle* of plants, to the colouring matter of which the chemico-botanist has given the name of *chromule*. The general hue of green, as employed by Nature in the vegetal world, is a compound of blue, or gray, and citrine, according to its situation in the fundamental scale of colours: the gayer compounds of blue and yellow she reserves for the decoration of the animal creation, as in birds, shells, insects, and fossils.

The principal discord of green is blue; and when they approximate or accompany each other, they require to be resolved by the apposition of warm colours; and it is in this way that the warmth of distance and the horizon reconcile the azure of the sky with the greenness of a landscape. Its less powerful discord is yellow, which requires to be similarly resolved by a purple-red, or its principles. In its tones green is cool or warm, sedate or gay, either as it inclines to blue or to yellow; yet it is in its general effects cool, calm, temperate, and refreshing; and, having little power in reflecting light, is a retiring colour, and readily subdued by distance: for the same reasons it excites the retina less than most colours, and is cool and grateful to the eye. As a colour individually, green is eminently beautiful and agreeable, but it is more particularly so when contrasted with its compensating colour, red; and they are the most generally attractive of all colours in this respect. They are hence powerful and effective colours on the feelings and passions, and require therefore to be subdued or toned to prevent excitement and to preserve the balance of harmony in painting.

The general powers of green, as a colour, associate it with the ideas of vigour and freshness; and it is hence symbolical of youth, the spring of life being analogous to the spring of the year, in which nature is surprisingly diffuse of this colour in all its freshness, luxuriance, and variety; soliciting the eye of taste, and well claiming the attention of the landscape-painter, according to the following judicious remarks of one of the most eminent of this distinguished class of British artists. "The autumn only is called the painter's season, from the great richness of the colours of the dead and decaying foliage, and the peculiar tone and beauty of the skies; but the spring has, perhaps, more than an equal claim to his notice and admiration, and from causes not wholly dissimilar,—the great variety of tints and colours of the living foliage, accompanied by their flowers and blossoms. The beautiful and tender hues of the young leaves and buds are rendered more lovely by being contrasted, as they now are, with the sober russet browns of the stems from which they shoot, and which still show the drear remains of the season that is past."—REMARKS ON LANDSCAPES CHARACTERISTIC OF ENGLISH SCENERY, BY J. CONSTABLE, ESQ. R.A.

Verdure is also the symbol of hope, which, like the animating greenness of plants, leaves us only with life: it is also emblematical of immortality, and the figure of old Saturn or Time is crowned with evergreen. This colour denotes also memory, and affords a great number of epithets and metaphors, colloquial as well as rhetorical. Plenty is personified in a mantle of green. In mythological subjects it distinguishes the draperies of Neptune, the Naiades, and the Dryades; and, from being a general garb of nature, perhaps, has been held to be a sacred or holy colour.

It is thus that colours lead ideas by association and analogy, and excite sentiments naturally in the manner we have so repeatedly alluded to already, in drawing attention to the powers of expression in colours; an attention of more importance than generally supposed in the practice of the pencil, and in spreading the charm of disguised art over its productions; and it is thus also that colouring moves the affections in the manner of certain chords in music, upon relations grounded so deeply in our nature, that a universal comprehension is perhaps requisite for their interpretation.

Of these powers of colours, which, when judiciously managed, are capable of producing or heightening sentiment, the poet has not failed to profit, of which we have already adduced many instances; and, in the absence of the more palpable illustrations of the powers and properties of green, by examples in nature and painting, the following are in point from the poets, expressing—

Youth, vigour, freshness, hope, &c.

——— My salad days,
When I was *green* in judgment, *cold in blood*.
SHAKSPEARE.

While *virgin Spring*, by Eden's flood,
Unfolds her *tender mantle green*.
BURNS.

And loud he sung agen the sunny shene;
O Maye, with all thy *flowres* and thy *grene*,
Right welcome be thou, faire, freshe Maye.
CHAUCER'S KNIGHT'S TALE, v. 1511.

But with your presence cheer'd, they cease to mourn,
And walks wear *fresher green* at your return.
DRYDEN.

If I have any where said *a green old age*, I have Virgil's authority: " *Sed crudâ Deo viridisque senectus.*"—DRYDEN.

The *green* stem grows in stature and in size,
But only feeds with *hope* the farmer's eyes.
ID. after OVID'S MET. lib. xv.

For May,—" Sweet Month! the groves *green* liveries wear,
If not the first, the fairest of the year;
For thee the Graces lead the dancing Hours,
And Nature's ready pencil paints the flowers."

You may be jogging while your boots are *green*.
SHAKSPEARE.

Green is indeed the colour of lovers.
IDEM.

Phyllidis adventu nostræ nemus omne *virebit*.
VIRGIL, Ecl. vii. 59.

Our griefs are *green*.
SHAKSPEARE.

Was the *hope* drunk
Wherein you dress'd yourself? Hath it slept since?
And wakes it now to look so *green and pale?*
IDEM.

Accordance with light and colours, &c. :—

Haste, haste, ye Naiads! with attractive art
New charms to every native grace impart:
With opening *flowrets* bind your *sea-green* hair.
ADDISON, AFTER STATIUS.

My mistaking eyes,
That have been so *bedazzled with the sun*,
That every thing I look on seemeth *green*.
SHAKSPEARE.

The *violets* now
That strew the *green* lap of the *new-come spring*.
IDEM.

Strike, louder strike th' ennobling strings
To those whose merchant sons were kings;
To him who deck'd with *pearly* pride,
For Adria weds his *green-hair'd* bride.
COLLINS.

Miranda! mark, where shrinking from the gale,
  Its silken leaves yet moist with early dew,
That *faint fair flower, the lily of the vale*,
  Droops its meek head, and looks, methinks, like you!
Wrapp'd in the *shadowy veil of tender green*,
  Its *snowy* bells a soft perfume dispense,
And bending, as reluctant to be seen,
  In simple loveliness it soothes the sense.
CHARLOTTE SMITH.

Seest how fresh my flowers been spread,
Dyed in *lily-white* and *crimson-red*,
With leaves *ingrain'd in lustie green*,
Colours meet to cloathe a maiden queen.
SPENSER'S SHEPHERD'S CALLENDER, FEB.

Discordance :—

> O beware, my Lord, of *jealousy*;
> It is the *green-eyed* monster, which doth mock
> The meat it feeds on.
>
> OTHELLO, Act III. Sc. 3.

> And what's a life? the flourishing array
> Of the proud summer-meadow, which to-day
> Wears her *green plush*, and is to-morrow *hay*.
>
> QUARLES, Emb. 13. B. 3.

> How all the other passions fleet to air,
> As doubtful thoughts, and rash-embraced despair,
> And shudd'ring fear, and *green-eyed jealousy!*
>
> SHAKSP., MER. OF VENICE.

Variety of contrast, &c. :—

> Such *crimson tempest* should bedrench
> The *fresh green* lap of *fair* King Richard's land.
>
> SHAKSPEARE.

> Britannia's genius bends to earth,
> And mourns the fatal day;
> While stain'd with *blood*, he strives to tear
> Unseemly from his *sea-green* hair
> The *wreaths of cheerful May*.
>
> COLLINS.

> That yon *green* boy shall have no sun to ripe
> The *bloom* that promiseth a mighty fruit.
>
> SHAKSPEARE.

> In a vault
> Where *bloody* Tybalt, yet but *green* in earth,
> Lies fest'ring in his *blood*.
>
> ROM. AND JULIET, Act IV. Sc. 3.

The entire import of the colours to the sentiment lies in general wider in context than the passages which we have extracted, with all possible brevity, from the poets;—and to analyse these critically would lead into a needlessly wide discussion, since this want will be easily supplied by the intelligent reader. This applies particularly to the colour green, which is as principal in poetry as it is in nature and painting.

Milton paints with green and violet, in a sort of minor key, thus beautifully :—

> Sweet Echo, sweetest nymph, that liv'st unseen,
> Within thy aery shell,

By *slow Meander's margent green*,
    And in the *violet-embroider'd vale*,
    Where the love-lorn Nightingale
Nightly to thee her sad song mourneth well.
                                        COMUS.

In the following, the elements of green combine in the joint sentiment or expression of youth, freshness, joy, and animation :—

Fair laughs the morn, and soft the zephyr blows,
    While proudly riding o'er the *azure* realm
In gallant trim the *gilded* vessel goes,
    Youth at the prow, and Pleasure at the helm.
                                        GRAY.

And in the succeeding, the expression of green is cool, refreshing, shadowy, &c. :—

Here in the sultriest season let him rest
Fresh in the green beneath those aged trees ;
Here winds of gentlest wing will fan his breast,
From heaven itself he may inhale the breeze.
                            BYRON, CHILDE HAROLD.

        Guide my way
Through fair Lyceum's walks, the *green retreats*
Of Academus, and the Thymy vale,
Where oft, enchanted with Socratic sounds,
Ilissus pure devolved his tuneful stream.
                        AKENSIDE, PL. IMAG. l. 590.

Bear me some God to Baia's gentle seats,
Or cover me in *Umbria's green retreats*,
Where western gales eternally reside,
And *all the seasons lavish all their pride*.
                    ADDISON'S EPIST. TO LD. HALIFAX.

Now waft me from the *green* hill's side,
    Whose *cold turf hides the buried friend!*
                                        COLLINS.

The *green leaves* quiver with the *cooling* wind.
                    SHAKSPEARE, TIT. AND. A. II. S. 3.

The *zephyrs curl* the *green locks* of the plain.
                                        DRUMMOND.

The *green* mantle of the standing pool.
                                        SHAKSPEARE.

> I know each lane and every *alley green*,
> Dingle, or bushy dell of this wild wood,
> And every bosky bourn.
>
> COMUS.

> The birds,
> Who all things now behold more *fresh and green*,
> After a night of storm so ruinous,
> Clear'd up their choicest notes.
>
> IBID.

The number of pigments of any colour is in general proportioned to its importance; hence the variety of greens is very great, though their classes are not very numerous. The following are the principal:—

I. MIXED GREENS. Green being a compound of *blue* and *yellow*, pigments of these colours may be used to supply the place of green pigments, by compounding them in the several ways of working—by mixing, glazing, hatching, or otherwise blending them in the proportions of the various hues required. The fine nature-like greens, which have lasted so well in some of the pictures of the Italian schools, appear to have been compounded of ultramarine and yellow. Whatever pigments are employed on a picture in the warm yellow hues of the foreground, and blue colouring of the distance and sky, are advantageous for forming the greens in landscape, &c., because they harmonize better both in colouring and chemically, and impart homogeneity to the whole,—which is a principle conducive to a fine tone and durability of effect; and this is a principle which applies to all mixed colours. In compounding colours, it is desirable not only that they should agree chemically, but that they should also have, as much as may be, the same degree of durability; and in these respects Prussian or Antwerp blue and gamboge form a judicious, though not extremely durable, compound, similar to *Varley's green, Hooker's green*, &c.

There is a green pigment of this kind prepared in Rome, of which the late President of the Royal Academy brought home a quantity, the modern substitute probably of the *Italian green* above mentioned, but wanting its durability, as it becomes blue in fading, and appears to be a mixture of Prussian blue and Dutch or Italian pink. See Cobalt Greens, Chrome Greens, and Prussian Green.

II. TERRE-VERTE. True Terre-Verte is an ochre of a bluish green colour not very bright, in substance moderately hard, and smooth in texture. It is variously a bluish or gray coaly clay combined with yellow oxide of iron or yellow ochre. Although not a bright, it is a very durable pigment, being unaffected by strong light and impure air, and combining with other colours without injury. It has not much body, is semitransparent, and dries well in oil. There are varieties of this pigment; but the green earths which have copper for their colouring matter are, although generally of brighter colours, inferior in their other qualities, and are not true terre-vertes.

The greens called *Verona green*, and *Verdetto, or holy green*, are similar native pigments of a warmer colour. These greens are found in the Mendip Hills, France, Italy, and the Island of Cyprus.

III. CHROME GREENS, commonly so called, are compound pigments, of which chrome yellow is the principal colouring substance. These are also called *Brunswick green*, &c. and are compounds of chromate of lead with Prussian and other blue colours, constituting fine greens to the eye, suitable to some of the ordinary purposes of mechanic art; but for obvious reasons before given are unfit for fine art. See Chrome Yellow. There is, however, a true chrome green, or *Native green*, the colouring matter of which is the pure oxide of chrome, and, being free from lead, is durable both against the action of the sun's light and impure air. It is of various degrees of transparency or opacity, and of several hues more or less warm or cool, which are all rather fine than brilliant greens, and afford pure and durable tints. True Chrome greens neither give nor receive injury from other pigments, and are eligible for either water or oil painting, in the latter of which they dry rapidly. The green of the Definitive Scale, Pl. 1. fig. 3. is of this latter kind.

IV. COBALT GREENS. There are two pigments of this denomination, the one a compound of cobalt blue and chromic yellow, which partakes of the qualities of those pigments, and may be formed on the palette,—the other, an original pigment prepared immediately from cobalt, which is of a pure but not very powerful green colour, and durable both in water and oil, in the latter of which it dries well.

V. 1. COPPER GREEN is the appellation of a class rather than of an individual pigment, under which are comprehended *Verdigris, Verditer, Malachite Mineral green, Green bice, Scheele's green, Schweinfurt* or *Vienna green*,

Emerald green, true *Brunswick green, green Lake, Mountain green, African green, French green, Saxon green, Persian green, Patent green, Marine green, Olympian green,* &c.; and old authors mention others under the names of individuals who prepared them, such are Verde de Barildo, &c.

The general characteristics of these greens are, brightness of colour, well suited to the purposes of house-painting, but not adapted to the modesty of nature in fine art. They have considerable permanence, except from the action of damp and impure air, which ultimately blacken them, to which shade they have also a tendency by time. They have a good body, and dry well in oil, but, like the whites of lead, are all deleterious substances. We will particularize the principal sorts:—

2. VERDIGRIS, or *Viride Æris,* is of two kinds, common or impure, and crystallized or *Distilled Verdigris,* or more properly refined verdigris. They are both acetates of copper, of a bright green colour inclining to blue. They are the least permanent of the copper greens, soon fading as water-colours by the action of light, &c. and becoming first white and ultimately black by damp and foul air. In oil verdigris is durable with respect to light and air, but moist and impure air change its colour, and cause it to effloresce or rise to the surface through the oil. It dries rapidly, and might be useful as a siccific with other greens or very dark colours. In varnish it stands better, but is not upon the whole a safe or eligible pigment, either alone or compounded. Vinegar dissolves it, and the solution is used for tinting maps, &c.

3. GREEN VERDITER is the same in substance as blue verditer, which is converted into green verditer by boiling. This pigment has the common properties of the copper greens above mentioned, and is sometimes called *Green Bice.*

4. EMERALD GREEN is the name of a new copper green upon a terrene base. It is the most vivid of this tribe of colours, being rather opaque and powerfully reflective of light, and appears to be the most durable pigment of its class. Its hue is not common in nature, but well suited for gems or glazing upon. It works well in water, but difficultly in oil, and dries badly therein.

5. MINERAL GREEN is the commercial name of *green lakes,* prepared from the sulphate of copper. These vary in hue and shade, have all the properties before ascribed to copper greens, and afford the best common greens, and, not being liable to change of colour by oxygen and light, stand the weather well, and are excellent for the use of the house-painter, &c.; but are less eligible in the nicer works of fine art, having a tendency to darken by time and foul air.

6. MOUNTAIN GREEN is a native carbonate of copper, combined with a white earth, and often striated with veins of mountain blue, to which it bears the same relation that green verditer does to blue verditer, nor does it differ from these and other copper greens in any property essential to the painter. The *Malachite*, a beautiful copper ore, employed by jewellers, is sometimes called mountain green, and *Green bice* is also confounded therewith, being similar substances and of similar use as pigments.

VI. SCHEELE'S GREEN is a compound oxide of copper and arsenic, named after the justly celebrated chemist who discovered it. It is variously of a beautiful light warm green colour, opaque, permanent in itself and in tint with white lead, but must be used cautiously with Naples yellow, by which it is soon destroyed. *Schweinfurt green* is the name of a fine preparation of the same kind. Both these pigments are less affected by damp and impure air than the simple copper greens, and are therefore in these respects rather more eligible colours than copper greens in general.

VII. PRUSSIAN GREEN. The pigment celebrated under this name is an imperfect prussiate of iron, or Prussian blue, in which the yellow oxide of iron superabounds, or to which yellow tincture of French berries has been added, and is not in any respect superior as a pigment to the compounds of Prussian blue and yellow ochre.

VIII. SAP GREEN, or *Verde Vessie*, is a vegetal pigment prepared from the juice of the berries of the buckthorn, the green leaves of the woad, &c. It is usually preserved in bladders, and is thence sometimes called *Bladder Green;* when good it is of a dark colour and glossy fracture, extremely transparent, and of a fine natural green colour. Though much employed as a water-colour without gum, which it contains naturally, it is a very imperfect pigment, disposed to attract the moisture of the atmosphere, and to mildew; and, having little durability in water-colour painting, and less in oil, it is not eligible in the one, and is totally useless in the other.

Similar pigments, prepared from coffee-berries, and called *Venetian and emerald greens*, are of a colder colour, very fugitive, and equally defective as pigments.

IX. INVISIBLE GREEN. See *Olive* Pigments.

CHAP. XIV.

OF PURPLE.

> Over his lucid arms
> A military vest of *purple* flow'd
> Livelier than *Melibœan*, or the grain
> Of *Sarra*, worn by kings and heroes old.
>
> MILTON.

PURPLE, the third and last of the secondary colours, is composed of *red* and *blue*, in the proportions of five of the former to eight of the latter, which constitute a perfect purple, or one of such a hue as will neutralize, and best contrast a perfect yellow in the proportions of thirteen to three, either of surface or intensity. It forms, when mixed with its co-secondary colour green, the tertiary colour *olive;* and when mixed with the remaining secondary orange, it constitutes in like manner the tertiary colour *russet*. It is the coolest of the three secondary colours, and the nearest also in relation to *black* or shade; in which respect, and in never being a warm colour, it resembles blue. In other respects also purple partakes of the properties of blue, which is its archeus or ruling colour; hence it is to the eye a retiring colour, which reflects light little, and declines rapidly in power in proportion to the distance at which it is viewed, and also in a declining light. It is owing to its being the mean between black and blue that it becomes the most retiring of all positive colours. Nature employs this hue beautifully in landscape, as a sub-dominant, in harmonizing the broad shadows of a bright sunshine ere the light declines into deep orange or red. Girtin, who saw Nature as she is, and painted what he saw, delighted in this effect of sun-light and shadow; but when purple is employed as a ruling colour in flesh, or otherwise, its effect is in general too cold, or verges on *ghastliness,* and is to be as much avoided as the opposite extreme of viciousness in colouring stigmatized as *foxiness.*

Yet, next to green, purple is the most generally pleasing of the consonant colours; and has been celebrated as a regal or imperial colour, as much perhaps from its rareness in a pure state, as from its individual beauty. It is probable, nevertheless, that the famed Tyrian purple was nearer to the rose, or red, than the purple of the moderns, in which inclination of hue this colour takes the names of *crimson,* &c., as it does those of *violet, lilac,* &c. when it inclines toward its other constituent, blue; which latter colour it serves to mellow, or follows well into shade.

The contrast, or harmonizing colour of purple, is yellow on the side of light and the primaries; and it is itself the harmonizing contrast of the tertiary *citrine* on the side of shade, and less perfectly so of the semi-neutral *brown.*

As purple, when inclining toward redness, is a regal and pompous colour, it has been used in mythological representations to distinguish the robe of Jupiter the king of Gods, and in general also as a mark of sacerdotal superiority: in its effects on the mind it partakes principally, however, of the powers of its archeus, or ruling colour, blue, and is hence a highly poetical colour, stately, dignifying, sedate, and grave; soothing in its lights, and saddening in its shades: accordingly it contributes to these sentiments under the proper management of the painter and the poet, as it does also popularly in its use in court mournings, and other circumstances of state: hence the poets sing of "*purple* state."

The rhapsodists of Greece often used to recite in a theatrical manner, not only with proper gestures, but in *colours* suitable to their subject; and when they thus acted the Odyssey of Homer, were dressed in a *purple-coloured* robe, ἁλιουργῷ, to represent the sea-wanderings of Ulysses: but when they acted the Iliad they wore one of a *scarlet colour,* to signify the bloody battles described in that poem. Upon their heads they wore a crown of *gold,* and held in their hands a wand made of the *laurel tree,* which was supposed to have the virtue of exciting poetic raptures. — See Sydenham on the Io of Plato, note 8. Eustath. on the Iliad, B. 1. and the Scholiast on Hesiod, Theog. vs. 50.

Of the various expression of purple, and its hues, we have the following examples from the poets:

> The *pale violet's dejected hue.*
>   AKENSIDE.

Death dyes the purple seas with gore.
THOMSON.

Shall we build to *the purple of pride?*
HERBERT KNOWLES.

Flowers of this purple dye,
Hit with Cupid's archery.
SHAKSPEARE.

Reclining soft in blissful bowers,
Purpled sweet with springing flowers.
FENTON.

Aurora had but newly chased the *night*
And *purpled* o'er the sky.
DRYDEN.

Bacchus, that first from out the *purple grape*
Crush'd the sweet poison of misused wine.
MILTON.

Lest from his hands the *purple* reins should slip.
SYMONIDES.

In the following, the poet employs purple in accordance with white and light, &c.

Not with more glories, in th' etherial plain,
The *sun first rises o'er the purpled main,*
Than, issuing forth, the rival of his beams
Launch'd on the bosom of the *silver Thames.*
Fair nymphs, and well-drest youths around her shone,
But every eye was fix'd on her alone.
On her *white breast* a sparkling cross she wore,
Which Jews might kiss, and Infidels adore.
POPE'S RAPE OF THE LOCK, Cant. II.

When next *the sun his rising light* displays,
And *gilds* the world below with *purple rays.*
DRYDEN'S VIRGIL.

Discordantly,—

Oft came Edward to my side,
With *purple falchion, painted to the hilt*
In blood of those that had encounter'd him.
SHAKSPEARE.

OF PURPLE. 135

>   He is come to ope
>   The *purple* testament of *bleeding* war.
>   SHAKSPEARE.

Contrasted,—

>   The lake return'd in chasten'd gleam
>   The *purple cloud*, the *golden beam*.
>   WALTER SCOTT.

>   Here Love his *golden* shafts employs; here lights
>   His constant *lamps*, and waves his *purple wings*.
>   MILTON.

>   When the glad harvest waves with *golden grain*,
>   And thirsty meadows drink the pearly rain,
>   On the proud vine her *purple gems* appear,
>   And smiling fields rejoice and hail the pregnant year.
>   PITT.

>   Aurora now, in radiant *purple* drest,
>   Shone from the portals of the *golden* east.
>   HOOLE'S TASSO.

>   Arraying with reflected *purple* and *gold*
>   The clouds that on his western throne attend.
>   MILTON.

By Composition,—

>   The Gods, who all things see, this same beheld,
>   And, pitying this paire of lovers true,
>   Transformed them there lying on the field
>   Into one flower that is both *red and blue*;
>   It first grows *red*, and then to *blue* doth fade,
>   Like *Astrophel*, which thereunto was made.
>   SPENSER, COL. CLO.

and in accordance with shade, &c.

>   In *darkness*, and with danger compass'd round,
>   And *solitude*, yet not alone, while thou
>   Visit'st my slumbers *nightly*, or when *morn*
>   *Purples the east*; still govern thou my song,
>   Urania!
>   IBID.

As the extreme primaries, blue and yellow, when either compounded or opposed, afford, though not the most perfect harmony, yet the most pleasing consonance of the primary colours; so the extremes, purple and orange, afford the most pleasing of the secondary consonances; and this analogy extends also to the extreme tertiary and semi-neutral colours, while the mean or middle colours afford the most agreeable contrasts or harmonies. This general feeling has evidently been that of the poet, with whom it is a licence to purple and gild without reserve, and with whom "to purple" often means to paint or colour in the abstract. Neither nature nor the painter's art is, however, so profuse of this colour, and purple pigments are rare, of which the following are the few that merit attention. Purple pigments lie under a peculiar disadvantage as to apparent durability and beauty of colour, owing to the neutralizing power of yellowness in the grounds upon which they are laid, as well as to the general warm colour of light, and the yellow tendency of almost all vehicles and varnishes, by which this colour is subdued.

I. MIXED PURPLES. Purple being a secondary colour, composed of *blue and red*, it follows of course that any blue and red pigments, which are not chemically at variance, may be used in producing mixed purple pigments of any required hue, either by compounding or grinding them together ready for use, or by combining them in the various modes of operation in painting. In such compounding, the more perfect the original colours are, the better in general will be the purple produced. In these ways *ultramarine* and the *rose colours of madder* constitute excellent and beautiful purples, which are equally permanent in water and oil, in glazing, or in tint, whether under the influence of the oxygenous or the hydrogenous principles of light and impure air, by which colours are subject to change. The blue and red of cobalt and madder afford also good purples. Some of the finest and most delicate purples in antient paintings appear to have been similarly compounded of *ultramarine and vermilion*, which constitute tints equally permanent, but less transparent than the above. Facility of use, and other advantages, are obtained at too great a sacrifice by the employment of perishable mixtures, such as are the carmines and lakes of cochineal with *indigo and other blue colours*.

II. GOLD PURPLE, or *Cassius's Purple Precipitate*, is the compound oxide which is precipitated upon mixing the solutions of gold and tin. It is not a bright, but a rich and powerful colour, of great durability, varying in hue from deep crimson to a murrey or dark purple, and is principally used in miniature, and may be employed in enamel-painting. It works well in water, and is an excellent though expensive pigment, but not much used at present, as the madder purple is cheaper, and perfectly well supplies its place.

III. MADDER PURPLE, *Purple Rubiate*, or *Field's Purple*, is a very rich and deep carmine, prepared from madder. Though not a brilliant purple, its richness, durability, transparency, and superiority of colour, have given it the preference to the purple of gold preceding, and to burnt carmine. It is a pigment of great body and intensity; it works well, dries and glazes well in oil, and is pure and permanent in its tints. It neither gives nor sustains injury from other colours, and is in every respect a very perfect and eligible pigment.

There is a lighter and brighter sort, which has all the properties of the above with less intensity of colour.

IV. BURNT CARMINE is, according to its name, the carmine of cochineal partially charred till it resembles in colour the purple of gold, for the uses of which in miniature and water-painting it is substituted, and has the same properties except its durability; of which quality, like the carmine it is made from, it is deficient, and therefore in this important respect is an ineligible pigment. A durable colour of this kind may, however, be obtained by burning *madder carmine* in a cup over a spirit lamp, or otherwise, stirring it till it becomes of the hue or hues required.

V. PURPLE LAKE. The best purple lake so called is prepared from cochineal, and is of a rich and powerful colour, inclined to crimson. Its character as a pigment is that of the cochineal lakes already described. It is fugitive both in glazing and tint; but, used in considerable body, as in the shadows of draperies, &c. it will last under favourable circumstances a long time. Lac lake resembles it in colour, and may supply its place more durably, although not perfectly so.

VI. LAC LAKE. See Red Lakes.

VII. PURPLE OCHRE, or MINERAL PURPLE, is a dark ochre,

native of the Forest of Dean in Gloucestershire. It is of a murrey or chocolate colour, and forms cool tints of a purple hue with white. It is of a similar body and darker colour than *Indian red,* which has also been classed among purples, but in all other respects it resembles that pigment.

CHAP. XV.

OF THE TERTIARY COLOURS.

OF CITRINE.

> His nose was high, his eyen bright *citrin*.
> CHAUCER'S KNIGHT'S TALE.

CITRINE or *citron* is the first of the tertiary class of colours, or ultimate compounds of the primary triad, *yellow*, *red*, and *blue;* in which *yellow* is the archeus or predominating colour, and blue the extreme subordinate; for citrine being an immediate compound of the secondaries, *orange* and *green*, of both which *yellow* is a constituent, the latter colour is of double occurrence therein, while the other two primaries enter singly into the composition of citrine,—its mean or middle hue comprehending eight blue, five red, and six yellow, of equal intensities.

Hence citrine, according to its name, which is the name of a class of colours, and is used commonly for a dark yellow, partakes in a subdued degree of all the powers of its archeus yellow; and, in estimating its properties and effects in painting, it is to be regarded as participating of all the relations of yellow. By some this colour is called brown, as almost all broken colours are. The harmonizing contrast of citrine is a *deep purple;* and it is the most advancing of the tertiary colours, or nearest in its relation to light. It is variously of a tepid, tender, modest, cheering character, and alike expressive of these qualities in pictorial and poetic art. In nature, citrine begins to prevail in landscape before the other tertiaries, as the green of summer declines; and as autumn advances it tends toward its orange hues, including the colours called aurora, chamoise, and others before enumerated under the head of Yellow.

OF THE TERTIARY COLOURS.—OF CITRINE.

To understand and relish the harmonious relations and expressive powers of the tertiary colours, requires a cultivation of perception and a refinement of taste to which study and practice are requisite. They are at once less definite and less generally evident, but more delightful,—more frequent in nature, but rarer in common art, than the like relations of the secondaries and primaries; and hence the painter and the poet afford us fewer illustrations of effects less commonly appreciated or understood. To this a want of right distinctions, and consequently also of proper appellations, may have contributed; nevertheless, the tertiaries have not escaped the eye of the poet, though his allusions to them are mostly ambiguous, metonymous, or periphrastical, as in the following examples of citrine:—

In accordance with light and shade, &c.

> Unmuffle, *ye faint stars, and thou, fair moon*,
> That wont'st to love the traveller's benison;
> *Stoop thy pale visage through an amber cloud*,
> And disinherit Chaos, that reigns here
> *In double night of darkness and of shades.*
>
> MILTON.

> The grete Emetrius, the king of Inde,
> Upon a stede *bay*, trapped in *stele*,
> Covered with cloth of *gold* diapred wele,
> Came riding like the god of arms.
> * * * * *
> His crispe here like ringes was yronne,
> And that was *yelwe*, and glitered as the sonne;
> His nose was high; his eyen bright *citrin;*
> His lippes round; his coloure was *sanguin.*
>
> CHAUCER'S KNIGHT'S TALE, v. 2158.

> While *sallow autumn fills thy lap with leaves.*
>
> COLLINS.

> Awake! the *morning shines*, and the fresh field
> Calls us: we lose the prime, to mark how spring
> Our tender plants, *how blows the citron grove*,
> What drops the *myrrh*, and what the *balmy reed*,—
> *How Nature paints her colours, how the bee
> Sits on the bloom*, extracting liquid sweets.
>
> MILTON.

> And on *the tawny sands and shelves*
> Trip the pert fairies and the dapper elves.
>
> IDEM.

Contrasted, &c.

> No *ivory work* my halls infold,
> Nor arched ceilings *gleaming gold;*
> Nor bear Hymettian columns wrought,
> *The citron beams from Afric brought:*
> Nor high-born dames would I e'er see
> *The Spartan purple* weave for me.
>
> HORACE, CARM. XVIII. lib. II.

> Ceres in *citrine* vest smiles on his care,
> And *purple* Bacchus brings him welcome cheer.
>
> ANONYMOUS.

> The *tawny* lion panting to get free.
>
> MILTON.

> Sardinia too, renown'd for *yellow fields*,
> With Sicily her bounteous tribute yields.
> No lands a glebe of richer tillage boast,
> Nor waft more plenty to the Roman coast.
>
> ROWE'S LUCAN, lib. II.

> His *tawny* beard was th' equal grace
> Both of his wisdom and his face;
> The upper part thereof was *whey*,
> The nether, *orange mix'd with grey.*
>
> BUTLER, HUDIBRAS, Chap. I.

Original citrine-coloured pigments are not numerous, unless we include several imperfect yellows, which might not improperly be called citrines: the following are, however, the pigments best entitled to this appellation, though we know of no one that bears it:—

I. MIXED CITRINE. What has been before remarked of the mixed secondary colours is more particularly applicable to the tertiary, it being more difficult to select three homogeneous substances, of equal powers as pigments, than two, that may unite and work together cordially. Hence the mixed tertiaries are still less perfect and pure than the secondaries; and as their hues are of extensive use in painting, original pigments of these colours are proportionately estimable to the artist. Nevertheless there are two evident principles of combination, of which the artist may avail himself in producing these colours in the various ways of working; the one being that of combining two original secondaries,—e. g. *green and orange* in producing a *citrine;* the other, the uniting the three primaries in such a manner that *yellow* predominate in the case of citrine, and *blue and red* be subordinate in the compound.

These colours are, however, in many cases produced with best and most permanent effect, not by the intimate combination of pigments upon the palette, but by intermingling them, in the manner of nature, on the canvas, so as to produce the effect at a proper distance of a uniform colour. Such is the *citrine* colour of fruit and foliage; on inspecting the individuals of which we distinctly trace the stipplings of orange and green, or yellow, red, and green. Similar beautiful consonances are observable in the *russet* hues of foliage in the autumn, in which purple and orange have broken or superseded the uniform green of leaves; and also in the *olive* foliage of the rose-tree, produced in the individual leaf by the ramification of purple in green. Yet mixed citrines may be compounded safely and simply by slight additions, to an original brown pigment, of that primary or secondary colour which is requisite to give it the required hue.

II. BROWN PINK is a vegetal lake precipitated from the decoction of French berries, and dyeing woods, and is sometimes the residuum of the dyer's vat. It is of a fine rich transparent colour, rarely of a true brown; but being in general of an orange broken by green, it falls into the class of citrine colours, sometimes inclining to greenness, and sometimes toward the warmth of orange. It works well both in water and oil, in the latter of which it is of great depth and transparency, but dries badly. Its tints with white lead are very fugitive, and in thin glazing it does not stand. Upon the whole, it is more beautiful than eligible.

III. CITRINE LAKE is a more durable and better drying species of brown pink, prepared from the quercitron bark. The citrine of the definitive scale, Pl. i. fig. 3., is of this pigment.

IV. CASSIA FISTULA is a native vegetal pigment, though it is more commonly used as a medicinal drug. It is brought from the East and West Indies in a sort of cane, in which it is naturally produced. As a pigment it is deep, transparent, and of an imperfect citrine colour, inclining to dark green—diffusible in water, without grinding, like gamboge and sap-green: it is, however, little used as a pigment, and that only in water, as a sort of substitute for bistre; which see.

V. UMBER, commonly called *Raw Umber*, is a natural ochre, abounding with oxide of manganese, said to have been first obtained from antient Ombria, now Spoleto, in Italy;—it is found also in England, and in most parts of the world; but that which is brought from Cyprus, under the name of Turkish umber, is the best. It is of a brown-citrine colour, semi-opaque, has all the properties of a good ochre, is perfectly durable both in water and oil, and one of the best drying colours we possess. Although not so much employed as formerly, it is perfectly eligible according to its colour and uses.

Several browns, and other ochrous earths, approach also to the character of citrines; such are the terre de Cassel, &c. But in the mixed confusion of names, infinity of tones and tints, and variations of individual pigments, it is impossible to attain an unexceptionable or universally satisfactory arrangement;—we have therefore followed a middle and general course in distributing pigments under their proper heads.

CHAP. XVI.

OF RUSSET.

'Tis sweet and sad the latest notes to hear
Of *distant music dying on the ear;*
'Tis sweet to hear *expiring summer's sigh,*
Thro' *forests tinged with* RUSSET, wail and die.
JOANNA BAILLIE.

THE second or middle tertiary colour, *Russet,* like citrine, is constituted ultimately of the three primaries, *red, yellow, and blue;* but with this difference, that instead of yellow as in citrine, *red* is the archeus or predominating colour in russet, to which yellow and blue are subordinates; for *orange* and *purple* being the immediate constituents of russet, and red being a component part of each of those colours, it enters doubly into their compound in russet, while yellow and blue enter it only singly; the proportions of its middle hue being eight blue, ten red, and three yellow, of equal intensities. It follows that russet takes the relations and powers of a subdued red; and many pigments and dyes of the latter denomination are in strictness of the class of russet colours: in fact, nominal distinction of colours is properly only relative; the gradation from hue to hue, as from shade to shade, constituting an unlimited series, in which it is literally impossible to pronounce absolutely where any shade or colour ends and another begins; but which is capable nevertheless of being arbitrarily divided to infinity.

The harmonizing, neutralizing, or contrasting colour of russet, is a *deep green;*—when the russet inclines to orange, it is a *gray,* or subdued blue. These are often beautifully opposed in nature, being medial accordances, or in equal relation to light, shade, and other colours, and among the most agreeable to sense.

Russet, we have said, partakes of the relations of red, but moderated in every respect, and qualified for greater breadth of display in the colouring of nature and art; less so, perhaps, than its fellow-tertiaries in proportion as it is individually more beautiful, the powers of beauty being ever most effective when least obtrusive; and its presence in colour should be principally evident to the eye that seeks it,—not so much courting as courted.

Of the tertiary colours, it is that which has supplied most of the ornament of epithet and sentiment to the poet; and his application of it is remarkably analogous to its just uses in painting when applied for the purposes of expression, which in this colour is warm, complacent, solid, frank, and cheering; of which and of its accordances and contrasts, &c., the following may serve as illustrations from the poets, who often, according to a common acceptation, substitute the term brown for russet:

> The Doric dialect has *a sweetness in its clownishness*; like *a fair shepherdess in her country russet*.
> DRYDEN.

> By this *white* glove, (how *white* the hand, God knows!)
> Henceforth my wooing mind shall be express'd
> In *russet* yeas, and honest *kersey* noes.
> SHAKSPEARE.

> But look—*the morn, in russet mantle clad*,
> Walks o'er the dew of yon high eastern hill.
> IDEM, HAMLET, A. 1. S. 1.

> Straight mine eye hath caught new pleasures,
> While the landscape round it measures;
> *Russet lawns*, and *fallows gray*,
> Where the nibbling flocks do stray.
> MILTON'S L'ALLEGRO.

> Here, in *full light, the russet plains extend;*
> There, *wrapt in clouds, the bluish hills* ascend.
> POPE.

> Around my *ivied porch* shall spring
>   Each fragrant flower that drinks the dew;
> And Lucy, at her wheel, shall sing
>   In *russet* gown and apron *blue*.
> ROGERS.

> A Palmer poor, in homely *russet* clad.
> DRAYTON.

> Our summer such a *russet livery* wears
> As in a garment often dyed appears.
> DRYDEN.

Of the tertiary colours, russet is the most important to the artist; and there are many pigments under the denominations of red, purple, &c., which are of russet hues; but there are few true russets, and one only which bears the name: of these are the following:—

I. MIXED RUSSET. What has been remarked in the preceding chapter upon the production of mixed citrine colours, is equally applicable in general to the mixed russets: we need not therefore repeat it. By the immediate method of producing it materially from its secondaries, orange vermilion and madder purple afford a compound russet pigment of a good and durable colour. Chrome orange and purple lake yield a similar but less permanent mixture.

Many other less eligible duple and triple compounds of russet are obvious upon principle, and it may be produced by adding red in due predominance to some browns; but all these are inferior to the following original pigments:—

II. RUSSET RUBIATE, *Madder Brown,* or *Field's Russet,* is, as its names indicate, prepared from the *rubia tinctoria,* or madder-root. It is of a pure, rich, transparent, and deep russet; of a true middle hue between orange and purple; not subject to change by the action of light, impure air, time, nor mixture of other pigments. It has supplied a great desideratum, and is indispensable in water-colour painting, both as a local and auxiliary colour, in compounding and producing with yellow the glowing hues of autumnal foliage, &c., and with blue the beautiful and endless variety of grays in skies, flesh, &c. There are three kinds of this pigment, distinguished by variety of hue, russet, or *madder brown, orange russet,* and purple russet, or *intense madder brown;* which differ not essentially in their qualities as pigments, but as warm or cool russets, and are all good glazing colours. The last dries best in oil, the others but indifferently. The russet of the definitive scale, Pl. i. fig. 3. is of the second kind.

III. PRUSSIATE OF COPPER differs chemically from Prussian blue only in having copper instead of iron for its basis. It varies in colour from russet to brown, is transparent and deep, but, being very liable to change in colour by the action of light and by other pigments, has been very little employed by the artist.

There are several other pigments which enter imperfectly into, or verge upon, the class of russet, which, having obtained the names of other classes to which they are allied, will be found under other heads; such are some of the ochres and Indian red. Burnt carmine and Cassius's precipitate are often of the russet hue, or convertible to it by due additions of yellow or orange; as burnt Sienna earth and various browns are, by like additions of lake or other reds.

CHAP. XVII.

OF OLIVE.

> The water-nymphs that wont to sing and dance,
>   And for her garland *olive* branches bear,
> Now baleful boughs of cypress do advance:
>   The Muses that were wont *green bays* to wear
>   Now bringen bitter *elder branches sere:*
>   The fatal sisters eke repent
>   Her vital thread so soon was spent:
>   O heavy hearse,
>   Mourn now, my muse, now mourn with heavy cheare;
>   O careful verse.
>   SPENSER, SHEPHERD'S CALLENDER.

OLIVE is the third and last of the tertiary colours, and nearest in relation to shade. It is constituted, like its co-tertiaries, citrine and russet, of the three primaries *blue, red,* and *yellow,* so subordinated, that blue prevails therein; but it is formed more immediately of the secondaries *purple* and *green;* and, since blue enters as a component principle into each of these secondaries, it occurs twice in the latter mode of forming olive, while red and yellow occur therein singly and subordinately. *Blue* is therefore, in every instance, the archeus or predominating colour of olive; its perfect or middle hue comprehending SIXTEEN of blue to FIVE of red, and THREE of yellow; and it participates in a proportionate measure of the powers, properties, and relations of its archeus: accordingly, the antagonist, or harmonizing contrast of olive, is a *deep orange;* and, like blue also, it is a retiring colour, the most so of all the colours, being the penultimate of the scale, or nearest of all in relation to *black,* and last, theoretically, of the regular distinctions of colours. Hence its importance in nature and painting is almost as great as that of black: it divides the office of clothing and decorating the general face of nature with green and blue; with both which, as with black and grey, it enters into innumerable compounds and accordances, changing its name, as either hue predominates, into green, gray, ashen, slate, &c.: thus the olive hues of foliage are called green, and the purple hues of clouds are called gray, &c., for language is general only, and inadequate to the infinite particularity of nature.

This infinity, or endless variation of tint, hue and relation, of which the tertiaries are susceptible, and which actually occur in nature, give a boundless license to the revelry of taste, in which the genius of the pencil may display the most captivating harmonies of colouring, and the most chaste and delicate expressions; too subtle to be defined, too intricate to be easily understood, and often too exquisite to be felt by the untutored eye; resembling, in this respect, what is recorded of the *enharmonic* of ancient music: and although these effects abound in nature, they lie for the most part beyond the simpler distinctions of the poet, which suffice him nevertheless for natural and moral allusions of exquisite beauty:

> Of every sort which in that meadow grew
> They gathered some; the *violet pallid hue*,
>   The little dazie that at evening closes,
> The *virgin lillie*, and the *primrose* trew,
>   With store of *vermeil roses*
> To deck their bridegroom's posies.
>
> SPENSER, PROTHALMION.

> For lo, my love doth in herself contain
> All this world's riches that may far be found;
> If *saphyres*, lo—her *eyes be saphyres* plain;
> If *rubies*, lo—her *lips be rubies* sound;
> If *pearls*, her *teeth be pearls* both pure and round;
> If *ivory*, her *forehead ivory* weane;
> If *gold*, her *locks are finest gold* on ground;
> If *silver*, her *fair hands are silver sheene:*
>   But that which fairest is, but few behold,
>   Her mind adorned with virtues manifold!
>
> IDEM, SONNETS.

The course of Nature, in the display of ruling colours expressive of the seasons in the vegetal creation, ere she ventures on the tertiaries, is worthy of remark: her first blossoms are *white and colourless*, as instanced by the poet,—

> Already now the *snow-drop* dares appear,
> *The first pale blossom of th' unripen'd year.*

> Fair Flora's breath, by some transforming power,
> Hath changed an icicle into a flower;
> Its name and hue the scentless plant retains,
> And winter lingers in its icy veins.

To which succeed flowers of *pale yellow and orange* hues:—

> *Daffodils*,
> That come before the swallow dares, and take
> The winds of March with beauty; *violets, dim*,
> But sweeter than the lids of Juno's eyes,
> Or Cytherea's breath; *pale primroses*,
> That die unmarried, ere they can behold
> Bright Phœbus in his strength.
>
> SHAKSPEARE.

Then follows, in fullest glow, the season of reds and of roses:—

> When Nature, *prodigal of flowers*,
> Holds her own court 'mid *rosy bowers;*
> Where the soft radiant summer's sky
> Spreads its ethereal canopy,
> *Deepening while mellowing its hue*
> In its *intensity of blue*.
>
> MARY ANN BROWN.

Which latter colour is latest in the general train of flowers, with those of strongest dye, *rich purple and deep blue*. Thus the scale of colours corresponds in a general way with the natural course of the seasons, for it is not indeed pretended that the order of the former is thus absolutely distinguished in the latter; such is not the method of Nature, who always melodizes by imperceptible gradations, while she harmonizes by distinct contrasts and accordances, so that flowers of all colours, variously subordinated, bloom at various seasons: and when the seasons of flowers may be considered as past, Nature, as if she had no farther use for her fine colours, or willing to display her ultimate skill and refinement, lavishes the contents of her palette, not disorderly, but in multiplied relations over all vegetal creation, in those rich and beautiful accordances of broken and finishing colours with which autumn is decorated ere the year decays and declines into darkness.

TO THE SEASONS.

Vernal and grassy, vivid, holy powers,
Whose balmy breath exhales in *lovely flowers;*
All-coloured Seasons, rich increase your care,
Circling for ever, *flourishing and fair;*
Invested with a *veil of shining dew,*
A *flowery veil* delightful to the view;
Attending *Proserpine;* when, *back from night,*
The Fates and Graces lead her up to light;
When in a band harmonious they advance,
And joyful round her form the solemn dance:
With Ceres triumphing; and Love, divine,
Propitious come, and on our incense shine;
Give Earth a blameless store of fruits to bear,
And make *a painter's mystic art* your care.

TAYLOR'S HYMNS OF ORPHEUS.

To distinguish these latter combinations of colours, we have observed that language gives no aid; they are therefore beyond the poet's pen, and it is the privilege of the pencil alone to represent or imitate them with just effect: and this affords us a true explanation of the paucity of the tertiary in poetry; yet olive is sometimes employed dubiously, with regard to colour, as symbolical of peace, solitude, and shade. Thus Collins, in his " Ode to Liberty,"—

And lo, an humbler relic laid
In *jealous Pisa's olive shade!*

Again,—

Where *peace*, with ever-blooming *olive*, crowns
The gate, where Honour's liberal hands effuse
Unenvy'd treasures.

AKENSIDE, PLEASURES OF IMAGINATION, B. I. l. 518.

So also Milton,—

But he her fears to cease
Sent down the meek-eyed Peace;
She, crown'd with *olive green*, came softly sliding
Down through the *burning sphere,*
His ready harbinger,
With turtle wings the amorous clouds dividing.

ODE ON THE NATIVITY.

And again,—

> Athens, the eye of Greece, mother of Arts
> And Eloquence, native to famous wits,
> Or hospitable; in her sweet recess,
> City or suburb, *studious walks and shades.*
> See there the *olive grove* of Academe,
> Plato's retirement, where the *Attic bird*
> Trills her thick-warbled notes the summer long.
>
> MILTON'S PARADISE REGAINED, B. IV. l. 240.

As Olive is usually a compound colour both with the artist and mechanic, and as there is no natural pigment in use under this name, or of this colour, in commerce, there are few olive pigments. *Terre-vert*, already mentioned, is sometimes of this class, and several of the copper greens acquire this hue by burning. The following need only be noticed:—

I. MIXED OLIVE may be compounded in several ways; directly, by uniting *green* and *purple*, or by adding to *blue* a smaller proportion of *yellow* and *red*, or by breaking much blue with little orange. Cool black pigments, combined with yellow ochre, afford eligible olives. These hues are called *green* in landscape, and *invisible green* in mechanic painting. It is to be noted that, in producing these and other compound colours on the palette or canvas, those mixtures will most conduce to the harmony of the performance which are formed of pigments otherwise generally employed in the picture.

II. OLIVE GREEN. The fine pigment sold under this name, principally as a water-colour, is an arbitrary compound, or mixed green, eligible for its uses in landscape, sketching, &c.

III. BURNT VERDIGRIS is what its name expresses, and is an olive-coloured oxide of copper deprived of acid. It dries remarkably well in oil, and is more durable; and, in other respects, an improved and more eligible pigment than the original verdigris. Scheele's green affords by burning also a series of similar olive colours, which are as durable as their original pigment.

IV. OLIVE LAKE is a lake prepared from the green ebony, and is of considerable durability, transparency, and great depth, both in water and oil, in which latter vehicle it dries well. The olive of the definitive scale, Pl. I., is of this pigment.

CHAP. XVIII.

OF SEMI-NEUTRAL COLOURS.

OF BROWN.

> Kate, like the hazel twig,
> Is straight and slender; and *as brown in hue*
> *As hazel-nuts*, and sweeter than their kernels.
> SHAKSPEARE, TAMING OF THE SHREW.

As colour, according to the regular scale descending from *white*, properly ceases with the class of *olive*, the neutral *black* would here naturally terminate the series; but as, in a practical view, every coloured pigment, of every class or tribe, combines with black as it exists in pigments,—not as deepened, or lowered in tone only, but also as defiled in colour, or changed in class,—a new series or scale of coloured compounds arises, having black for their basis, which, though they differ not theoretically from the preceding order inverted, are, nevertheless, practically imperfect or impure; in which view, and as compounds of black, we have distinguished them by the term *semi-neutral*,* and divided them into three classes, corresponding to the natural relations preceding, and comporting with our common appellations and conceptions of colours, and the greater number of our natural and artificial pigments, between the extremes of the indefinite terms brown and gray. Inferior as the semi-neutrals are in point of colour, they comprehend, nevertheless, a great proportion of our most permanent pigments; and are, with respect to black, what *tints* are with respect to white, i. e. they are, so to call them, black tints, or shades.

* See Note D.

The first of the semi-neutrals, and the subject of the present chapter, is BROWN, which, in its widest acceptation, has been used to comprehend vulgarly every denomination of dark broken colour, and, in a more limited sense, is the rather indefinite appellation of a very extensive class of colours of warm or tawny hues. Accordingly we have browns of every denomination of colours except blue; thus we have yellow-brown, red-brown, orange-brown, purple-brown, &c.: but it is remarkable that we have, in this sense, no blue-brown, nor any other coloured-brown, in any but a forced sense, in which blue predominates; such predominance of a cold colour immediately carrying the compound into the class of gray, ashen, or slate-colour. Hence brown comprehends the hues called feuillemort, mort d'ore, dun, hazle, auburn, &c.; several of which we have already enumerated as allied to the tertiary colours.

The term *brown*, therefore, properly denotes a warm broken colour, of which *yellow* is a principal constituent: hence brown is in some measure to shade what yellow is to light, and warm or ruddy browns follow yellows naturally as shading or deepening colours. It is hence also that *equal quantities* of either the three primaries, the three secondaries, or the three tertiaries, produce variously a brown mixture, and not the neutral black, &c.; because no colour is essentially single, and warmth belongs to two of the primaries, but coldness to blue alone.

This tendency in the compounds of colours to run into brownness and warmth, is one of the general natural properties of colours, which occasions them to deteriorate or dirt each other in mixture: hence *brown* is synonymous with foul or defiled, in a sense opposed to *fair* and pure; and it is hence also that brown, which is the nearest of the semi-neutrals in relation to light, is to be avoided in mixture with light colours. It is nevertheless an example of the wisdom of the Author of nature that brown is rendered, like green, a prevailing hue, and in particular an earth colour, as a contrast which is harmonized by the blueness and coldness of the *sky*,—both these colours prevailing together, too, in hot climates.

This tendency will account also for the use of brown in harmonizing and toneing, and for the great number of natural and artificial pigments and colours we possess under this denomination: in fact, the failure to produce other colours chemically or by mixture is commonly productive of a brown, which, on the other hand, is the colour of dirt, and defiles all other colours. It was this fertility and abundance of brown that occasioned our great landscape-colourist Wilson, when an acquaintance went exultingly to inform him he had discovered a new brown, to check him, in his characteristic way, with—" *I'm sorry for it:—we have gotten too many of them already.*" Yet are fine transparent browns obviously very valuable colours. If red or blue be added to brown predominantly, it falls into the other semi-neutral classes, marrone or gray.

The wide acceptation of the term brown has occasioned much confusion in the naming of colours, since broken colours in which red, &c. predominate have been improperly called brown; and a tendency to red or hotness in browns obtains for them the reproachful appellation of *foxiness*. This term, brown, should therefore be confined to the class of semi-neutral colours, compounded of, or of the hues of, either the *primary yellow, the secondary orange, or the tertiary citrine, with a black pigment;* the general contrast or harmonizing colour of which will consequently be more or less purple or blue; and with reference to black and white, or light and shade, it is of the semi-neutrals the nearest in accordance with white and light.

Brown is a sober and sedate colour, grave and solemn, but not dismal, and contributes to the expression of strength, stability, and solidity,—vigour, warmth, and rusticity,—and in minor degree to the serious, the sombre, and the sad; not with the painter only, but also with the rhetorician and poet, with whom, nevertheless, many of the broken colours are yet " airy nothings " and " without a name."

Examples of the applications of brown to the purposes of sentiment abound with the poets. Milton employs this colour in the beginning of his " Monody of Lycidas " thus plaintively :

> Yet once more, *O ye laurels, and once more,*
> *Ye myrtles brown, with ivy never sere,*
> I come to pluck your berries harsh and crude,
> And with forced fingers rude,
> *Shatter your leaves before the mellowing year ;*
> For Lycidas is dead—.

And in the following, from an unknown hand, brown is thus beautifully associated with true feeling of the force of colouring :

> Go, mark in meditative mood where Autumn
> Steals o'er his woods with mellowing touch, like Time
> Ripening the tints of some delicious Claude.
> Mark where each elm hangs forth its *golden* bough,

Like that which Virgil's hero sought;* each oak
Fades picturesquely *brown;* each sycamore
Mantled with *ruddy* richness. 'Tis a scene
That o'er us sheds *the mild and musing calm
Of wisdom; breathes, as noblest bards have own'd,
Poetic inspiration; bids us taste
The lonely sweetness of a walk with her
By Milton wooed,* "*divinest Melancholy.*"
And wouldst thou go, unfeeling, and prefer
The gorgeous blaze of summer to the charm—
The dying charm of Autumn's farewell smile?
ANON.

It is apparent that in many instances the colours in which the poet clothes his figures are chosen purely for their expression, as in the following:

Satyrs and sylvan boys were seen
Peeping from forth their alleys green;
Brown Exercise rejoiced to hear,
And Sport leap'd up and seized his *beechen* spear.
COLLINS.

—— In *the dun air sublime.*
MILTON.

And see, the fairy valleys fade,
Dun night has veil'd the solemn view!
Yet once again, dear parted shade,—
Meek Nature's child, again adieu!
COLLINS.

Be mine the hut
That from the mountain's side
Views wilds, and swelling floods,
And *hamlets brown,* and dim-discover'd spires,
And hears their simple bell; and marks o'er all
Thy dewy fingers draw
The dusky veil.
IDEM, ODE TO EVENING.

The unsightly plain
Lies *a brown deluge.*
THOMSON.

* Æn. VI.

Byron justly and beautifully contrasts this hue thus:

> How throbs the pulse when first we view
> The eye that rolls in *glossy blue,*
> Or *sparkles black;* or mildly throws
> A beam *from under hazel brows.*

And Goldsmith thus:

> Low lies that house where *nut-brown* draughts inspired,
> Where *grey-beard* Mirth and smiling Toil retired.

Again, Mrs. Barbauld:

> But thou, O nymph! retired and coy,
> In what *brown hamlet* dost thou joy
>   To tell thy tender tale?
> The lowliest children of the ground,
> *Moss-rose and violet,* blossom round,
>   And *lily of the vale.*
> ODE TO CONTENT.

And again:

> O'er *his fair brow, the fairer for their shade,*
> Locks of the *warmest brown* luxuriant play'd.
> Blushing he bows—and gentle awe supplies
> Each flattering meaning to his downcast eyes;
> Sweet, serious, tender, those *blue* eyes impart
> A thousand dear sensations to the heart.
> MISS SEWARD.

Sir Walter Scott thus contrasts brown with fair:—

> Wreathed in its *dark-brown* rings her hair,
> Half-hid, Matilda's forehead *fair;*
> Half hid, and half reveal'd to view,
> Her full *dark eyes of hazel hue.*
> ROKEBY.

Shakspeare calls brown dissembling, perhaps in the sense of shadowing, as the Italians call burnt umber *falsalo:*—

> His very hair is of the *dissembling colour*—
> Something *browner* than Judas's.
> AS YOU LIKE IT.

And in the following it is accorded with shade, black, &c. :—

> To *arched walks of twilight groves*,
> And *shadows brown*, that Sylvan loves,
> Of pine or *monumental oak*.
> MILTON'S LYCIDAS.

> Not that our heads are,
> Some *brown, some black, some auburn, and some bald*,
> But that our wits are *so diversely coloured*.
> SHAKSPEARE.

> *Obscured* where highest woods, *impenetrable*
> *To stars or sun-light*, spread their *umbrage* broad,
> And *brown as evening*.
> MILTON.

> Bid this mild summer of my life rebloom ;
> Bid *every shade embrown* and *cloister gloom*.
> MASON.

> The *rustic youth, brown with meridian toil*,
> *Healthful and strong :* full as the summer *rose*,
> Blown by prevailing suns, *the ruddy maid* —.
> THOMSON.

> Where beasts with man divided empire claim,
> And the *brown Indian marks with murderous aim*.
> GOLDSMITH.

> Ambitious of the town,
> She left her wheel and robes of *country brown*.
> IDEM.

> The *fields, all iron, cast a gleaming brown*.
> MILTON.

The list of brown pigments is very long, and that of MIXED BROWNS literally endless, it being obvious that every warm colour mixed with black will afford a brown, and that equal portions of the primaries, secondaries, or tertiaries, will do the same ; hence there can be no difficulty of producing them by mixture when required, which is seldom, as there are many browns which are good and permanent pigments among the following :—

I. VANDYKE-BROWN. This pigment, hardly less celebrated than the great painter whose name it bears, is a species of peat or bog-earth of a fine, deep, semi-transparent brown colour. The pigment so much esteemed and used by Vandyke is said to have been brought from Cassel ; and this seems to be justified by a comparison of *Cassel-earth* with the browns of his pictures.* The Vandyke-browns in use at present appear to be terrene pigments of a similar kind, purified by grinding and washing over: they vary sometimes in hue and in degrees of drying in oil, which they in general do tardily, owing to their bituminous nature, but are good browns of powerful body, and are durable both in water and oil. The *Campania-brown* of the old Italian painters was a similar earth.

II. MANGANESE BROWN is an oxide of manganese, of a fine, deep, semi-opaque brown of good body, which dries admirably well in oil. It is deficient of transparency, but may be a useful colour for glazing or lowering the tone of white without tinging it, and as a local colour in draperies, dead-colouring, &c. It is a perfectly durable colour both in water and oil.

III. RUBENS'-BROWN. The pigment still in use in the Netherlands under this appellation is an earth of a lighter colour and more ochrous texture than the Vandyke-brown of the London shops: it is also of a warmer or more tawny hue than the latter pigment, and is a beautiful and durable brown, which works well both in water and oil, and much resembles the brown used by Teniers.

IV. CASSEL-EARTH, or, corruptly, *Castle-earth*. The true *terre de Cassel* is an ochrous pigment similar to the preceding, but of a brown colour, more inclined to the russet hue. In other respects it does not differ essentially from Rubens' and Vandyke-browns.

V. COLOGN-EARTH, incorrectly called *Cullen's-earth*, is a native pigment, darker than the two last, and in no respect differing from Vandyke-brown in its uses and properties as a colour. Similar earths abound in our own country.

VI. BURNT UMBRE is the fossil pigment called Umbre, burnt, by which it becomes of a deeper and more russet hue, and very drying in oil, in which it is employed as a dryer. It may be substituted for Vandyke-

* In the tribe of browns—in oil painting one of the finest earths is known, at the colour-shops, by the name of *Castle-earth or Vandyke's-brown.*—Gilpin's Essays on Picturesque Beauty, &c. Essay IV. p. 33.

brown, and is a perfectly durable and eligible pigment in water, oil, or fresco. The old Italians called it *falsalo*.

VII. BROWN OCHRE. See *Yellow Ochre*.

VIII. SPANISH BROWN, or *Tiver*. See *Red Ochre*.

IX. BONE BROWN and *Ivory Brown* are produced by torrefying, or roasting, bone and ivory till by partially charring they become of a brown colour throughout. They may be made to resemble the five first browns above by management in the burning; and, though much esteemed by some artists, are not perfectly eligible pigments, being bad dryers in oil, and their lighter shades not durable either in oil or water when exposed to the action of strong light, or mixed in tint with white lead. The palest of these colours are also the most opaque: the deepest are more durable, and most so when approaching black.

X. ASPHALTUM, called also *Bitumen, Mineral Pitch*, and *Jews' Pitch*, is a resinous substance rendered brown by the action of fire, natural or artificial. The substance employed in painting under this name is black and glossy like pitch, which differs from it only in having been less acted upon by fire, and in thence being softer. Asphaltum is principally used in oil-painting; for which purpose it is first dissolved in oil of turpentine, by which it is fitted for glazing and shading. Its fine brown colour, and perfect transparency, are lures to its free use with many artists, notwithstanding the certain destruction which awaits the work on which it is much employed, owing to its disposition to contract and crack by changes of temperature and the atmosphere; but for which it would be a most beautiful, durable, and eligible pigment. The solution of asphaltum in turpentine, united with drying oil by heat, or the bitumen torrefied and ground in linseed or drying oil, acquires a firmer texture, but becomes less transparent, and dries with difficulty.

A specimen of the native bitumen, brought from Persia by Lieutenant Ford, of which we made trial, had a powerful scent of garlic when rubbed. In the fire it softened without flowing, and burnt with a lambent flame; did not dissolve by heat in oil of turpentine, but ground easily as a pigment in pale drying oil, affording a fine deep transparent brown colour, resembling that of the asphaltum of the shops; and dried firmly nearly as soon as the drying oil alone. Asphaltum may be used as a permanent brown in water, and the native kind is also superior to the artificial for this purpose.

XI. MUMMY, or *Egyptian Brown*, is also a bituminous substance combined with animal remains, brought from the catacombs of Egypt, where liquid bitumen was employed three thousand years ago in embalming, in which office it has combined, by a slow chemical change, during so many ages with substances which give it a more solid and lasting texture than simple asphaltum; but in this respect it varies exceedingly, even in the same subject. Its other properties and uses as a pigment are the same as those of asphaltum, for which it is employed as a valuable substitute, being less liable to crack or move on the canvas. This also may be used, when ground, as a water-colour.

XII. ANTWERP BROWN is a preparation of asphaltum ground in strong drying oil, by which it becomes less liable to crack. See the two last articles. Bituminous coal, jet, and other bituminous substances, afford similar browns.

XIII. BISTRE is a brown pigment extracted by watery solution from the soot of wood-fires, whence it retains a strong pyroligneous scent. It is of a wax-like texture, and of a citrine-brown colour, perfectly durable. It has been much used as a water-colour, particularly by the old masters in tinting drawings and shading sketches, previously to Indian ink coming into general use for such purposes. In oil it dries with the greatest difficulty.

A substance of this kind collects at the back of fire-places in cottages where peat is the constant fuel burnt; which, purified by solution and evaporation, affords a fine bistre. Scotch bistre is of this kind. All kinds of bistre attract moisture from the atmosphere.

XIV. SEPIA, *Seppia, or Animal Æthiops*. This pigment is named after the sepia, or *cuttle-fish*, which is called also the *ink-fish* from its affording a dark liquid which was used as an ink by the ancients. From this liquid our pigment sepia, which is brought principally from the Adriatic, and may be obtained from the fish on our own coasts, is said to be obtained; and it is supposed that it enters into the composition of the *Indian ink* of the Chinese. Sepia is of a powerful dusky brown colour, of a fine texture, works admirably in water, combines cordially with other pigments, and is very permanent.

It is much used as a water-colour, and in making drawings in the manner of bistre and Indian ink; but is not used in oil, in which it dries very reluctantly.

XV. HYPOCASTANUM, or *Chestnut Brown*, is a *brown lake* prepared from the horse-chestnut; transparent and rich in colour, warmer than brown pink, and very durable both in water and oil; in the latter of which it dries moderately well.

XVI. MADDER BROWN. See *Russet Rubiate*, II. chap. xvi.

XVII. BROWN PINK. See Chap. xv. II.

XVIII. PRUSSIAN BROWN is a preparation of Prussian blue, from which the blue colouring principle has been expelled by fire, or extracted by an alkaline ley; it is an orange brown, of the nature and properties of Sienna earth.

XIX. BROWN INK. Various of these were used in sketching by Claude, Rembrandt, and many of the old masters; the principal of which were solutions of bistre and sepia. Less eligible preparations, which have faded or decayed, as common ink does, have also been employed; but other sketching inks may easily be produced, of durable colours and agreeable tints.

CHAP. XIX.

OF MARRONE.

We have adopted the term MARRONE for our second and middle semi-neutral, as univocal of a class of impure colours composed of black and red, black and purple, or black and russet pigments, or with black and any other denomination of pigments in which red predominates. It is a mean between the warm, broken, semi-neutral class of colours called *brown*, and the cold semi-neutral class of *grey*, or *ashen*. Marrone is practically to shade, what red is to light; and its relations to other colours are those of red, &c., when we invert or degradate the scale from black to white. It is therefore a following, or shading, colour of red and its derivatives; and hence its accordances, contrasts, and expressions agree with those of red degraded; hence red added to brown converts it into marrone if in sufficient quantity to predominate. In smaller proportions red gives to browns the denominations of bay, chestnut, sorrel, &c.

Owing to the instability and confusion of the nomenclature of colours, most of the colours and pigments of this class have been assigned to other denominations, as reds, browns, and purples—puce, pavonazzo, murrey, morello, chocolate, &c. (and the seasons of London bring us annually new names for broken colours from the dyer, few of which survive the ephemeral fashions which introduce them): hence pigments which belong properly to the present and other classes, have been arranged according to their names under other heads; such in the present instance are the ochres called purple-brown, mineral purple, dark cassius purple, dark Indian red, &c., which see. It is owing to this vagueness of nomenclature that the present and other denominations of broken colours have been unknown to or little used by the poet.

Marrone is a colour easily compounded in all its hues and shades by the mixture variously of red, black, and brown; but the following is the only pigment which bears the denomination:—

I. MARRONE LAKE is a preparation of madder of great depth, transparency, and durability of colour. It works well in water, glazes and dries in oil, and is in all respects a good pigment: as, however, its hues are easily given with other pigments, it has not been much used. There is a deeper kind, which has been called *purple-black*.

II. CARUCRU, or *Chica*, is a new pigment, of a soft powdery texture, and rich marrone colour, brought by Lieutenant Mawe from South America; for a portion of which we have been indebted to the kindness of Mr. Brockedon. It is said to be procured from a species of bignonia in the manner of indigo by the Indians of the interior of Guiana, and employed by their chiefs and higher orders as a fucus for the face, and as a sovereign remedy, topically applied, for the erysipelas.* Comparatively as a pigment, it resembles marrone lake in colour, and is equal in body and transparency to the carmine of cochineal, though by no means approaching it in beauty, or even in durability, fugitive as the latter pigment is. Exposed to the light of a window, even without sun, the colour of carucru is soon changed and destroyed, which defects alone render it unfit for fine art, whatever value it may be found to possess in dyeing or in medicine.

In its chemical affinities it very much resembles the best anotta, although it is redder in colour; and, if we may venture an opinion, it is but a finer species of that drug, and may be substituted for it in tinging lackers and varnishes, as it forms a rich orange tincture with spirit of wine. Its use as a rouge evinces a good eye in the Carib, with whose complexion it is better suited to harmonize, than the gaudy rouge prepared from carthamus or safflower, very injudiciously employed by the fairer beauties of Europe.

* See an article on this production by Dr. Hancock, Edin. N. Phil. Journ., No. xiv.

CHAP. XX.

OF GRAY.

*Down sunk the sun, the closing hour of day
Came onward, mantled o'er with dusky gray.*

PARNELL.

OF the tribe of semi-neutral colours, GRAY is the third and last, being nearest in relation of colour to black. In its common acceptation, and that in which we here use it, gray denotes a class of cool cinereous colours, faint of hue; whence we have blue grays, olive grays, green grays, purple grays, and grays of all hues, in which blue predominates; but no yellow or red grays, the predominance of such hues carrying the compounds into the classes of brown and marrone, of which gray is the natural opposite. In this sense the *semi-neutral* GRAY is distinguished from the *neutral* GREY, which springs in an infinite series from the mixture of the neutral *black* and *white*:—between *grays* and *greys*, however, there is no intermediate, since where *colour* ends in the one, *neutrality* commences in the other, and *vice versa*;—hence the natural alliance of the semi-neutral gray with black or shade; an alliance which is strengthened by the latent predominance of blue in the synthesis of black, so that in the tints resulting from the mixture of black and white so much of that hue is developed as to give apparent colour to the tints. This affords the reason why the tints of black and dark pigments are colder than their originals, so much so as in some instances to answer the purposes of positive colours; and it accounts in some measure for the natural blueness of the sky, though this is partly dependent, by contrast, upon the warm colour of sunshine to which it is opposed; for, if by any accident the light of nature should be rendered red, the colour of the sky would not appear purple in consequence, but green; or if the sun shone green the sky would not be green, but red inclined to purple; and so on of all colours, not according to the laws of composition in colours, but of contrast, since, if it were otherwise, the golden rays of the sun would render a blue sky green.

The *grays* are the natural cold correlatives, or contrasts, of the warm semi-neutral *browns;* and they are degradations of blue and its allies;—hence *blue* added to brown throws it into or toward the class of grays, and hence grays are equally abundant in nature and necessary in art; for the grays comprehend in nature and painting a widely diffused and beautiful play of retiring colours in skies, distances, carnations,' and the shadowings and reflections of pure light, &c. Gray is indeed the colour of space, and hence has the property of diffusing breadth in a picture, while it furnishes at the same time good connecting tints, or media, for harmonizing the general colouring: the grays are therefore among the most essential hues of the art, which yet must not be suffered injudiciously to predominate in cases where the subject or sentiment does not require it, so as to cast over the work the gloom or leaden dulness reprobated by Sir Joshua Reynolds; * although in solemn subjects they are wonderfully effective and proper ruling colours.

As blue is the archeus of all the colours which enter into the composition of grays, the latter partake of the relations and affections of blue, both with the painter and the poet. *Grave* sounds, like *gray* colours, are deep and dull, and there is a similarity of these terms in sound, signification, and sentiment, if even they are not of the same etymology: be this as it may, *gray* is almost as common with the poet, and in its colloquial use, as it is in nature and painting. The grays, like the other semi-neutrals, are sober, modest colours, contributing to the expression of gloom, sadness, frigidity, and fear,—the grave, the obscure, the spectral,—age, decrepitude, and death; bordering in these respects upon the powers of *black,* but aiding the livelier and more cheering expressions of other *colours* by diversity, connexion, and contrast, and partaking of the more tender and delicate influence belonging to *white,* as they approach it in their lighter tints. Upon the whole, it may be inferred as a general rule, that half of a picture ought to be of a neutral hue, to insure the harmony of the colouring, or at least that a balance of colour and neutrality is quite as essential to the best effect of a picture as a like balance of light and shade is, so universal is the reference of gray; and hence the frequent allusions to this colour by the poets thus variously:—

> Put your *torches* out—the gentle day,
> Before the wheels of Phœbus, round about
> Dapples the drowsy east with spots of grey.
>
> SHAKSPEARE.

* Reynolds's Works by Farrington, Notes, vol. III. p. 162.

For all was *blank, and bleak, and gray,*—
It was not night—it *was not day.*
BYRON, PRIS. OF CHILLON.

Cæcelia, that is *gray-eyed.*
CAMDEN.

Oh! how unseemly shows in *blooming youth*
Such *grey severity!*
MILTON, COMUS.

His hair just *grizzled* as in a *green old age.*
DRYDEN.

Gray-headed men and *grave* with warriours mixt.
MILTON.

Oft have I seen *a timely-parted ghost*
Of ashy semblance, meagre, pale, and bloodless.
SHAKSPEARE.

Though *grey*
Do something mingle with our *younger brown.*
IDEM, ANT. AND CLEOP., Act IV. Sc. 8.

The roses in thy lips and cheeks shall fade
To paly ashes; thy eye's windows fall.
IDEM, ROM. AND JULIET, Act IV. Sc. 1.

Gray-beard, thy love doth *freeze.*
IDEM.

Have I in conquest stretch'd mine arms so far
To be afraid to tell *gray-beards* the truth?
IDEM.

Her eyes are *gray as glass,* and so are mine.
IDEM.

The *grey-eyed morn* smiles on the frowning night,
Checkering the eastern clouds with streaks of light;
And *flecked darkness* like a drunkard reels
From forth day's pathway —.
IDEM, ROM. AND JULIET, Act II. Sc. 3.

Black spirits and white,
Blue spirits and grey,
Mingle, mingle, mingle—.
IDEM, MACBETH, Act IV. Sc. 1.

Our *green youth* copies what our *gray sinners* act.
DRYDEN.

Gray-headed infants.
IDEM.

They left me then when the *gray-hooded even'*,
Like a sad votarist in palmer's weed,
Rose from the hindmost wheels of Phœbus' wain.
                                            MILTON.

Now came still evening on, and *twilight gray*,
Had, *in her sober livery*, all things clad.
                                            ID. PAR. LOST.

Ye mists and exhalations, that now rise
From hill or streaming lake, *dusky or gray*,
Till *the sun paint your fleecy skirts with gold*,
In honour of the world's Great Author, rise.
                                            IDEM.

Thus pass'd the *night so foul*, till *morning fair*
Came forth with pilgrim steps, in *amice gray*.
                                            ID. PAR. REG.

I would not leave old Scotland's mountains *gray*,
  Her hills, her cots, her halls, her groves of pine,
Dark though they be; yon glen, yon broomy brae,
  Yon wild fox cleugh, yon eagle cliff's outline;
An hour like this—this *white* right hand of thine,
  And of thy *dark* eye such a gracious glance,
As I got now, for all beyond the line,
  And all the glory gain'd by sword or lance,
In gallant England, Spain, or olive vales of France.
                                            A. CUNNINGHAM.

And some will *mourn in ashes*, some *coal black*.
                                            SHAKSPEARE.

All music sleepe, where death doth lead the dance,
   And shepherds' wonted solace is extinct—
   The blue in black, the green in gray is tinct:
The *gaudy garlands* deckt her grave,
The *faded flowers* her corse embrave.
   O heavie hearse!
Mourn now my muse, now mourn with tears besprent.
   O careful verse!
   The mantled meadows mourne,
   Their *sundry colours tourne*.
                                            SPENSER'S SHEP. CAL. *Nov.*

There are several pigments of this class, which follow; and others might easily be found if required. They are also as easily compounded as they are useful and essential in painting:—

I. MIXED GRAYS are formed not only by the compounding of black and white, which yields *neutral greys*, and of black and blue, black and purple, black and olive, &c., which yield the *semi-neutral grays* of clouds, &c., but these may be well imitated by the mixture of russet rubiate, or madder browns, with blues, which form transparent compounds, which are much employed: Grays are, however, as above remarked, so easily produced, that the artist will in this respect vary and suit his practice to his purpose.

II. NEUTRAL TINT. Several mixed pigments of the class of gray colours are sold under the name of Payne's gray, neutral tint, &c. They were first employed by Cousins, and are, as we have been informed, now variously composed of sepia and indigo or other blues, with madder or other lakes, and are designed for water-colour painting only, in which they are found extremely useful. And here it may be proper to mention those other pigments, sold under the name of tints, which belong to no particular denomination of pigments; but being compounds, the result of the experience of accredited masters in their peculiar modes of practice, serve to facilitate the progress of their pupils, while they are eligible in a like view to other artists. Such are *Harding's and Macpherson's tints*, usually sold ready prepared in cakes and boxes for miniature and water painting. The latter of these two we know to be composed of pigments which associate cordially and with permanence, and may therefore be relied upon; nevertheless the artist will in general prefer a dependence upon his own skill for the production of his tints in painting.

III. ULTRAMARINE ASHES are the recrement of Lapis lazuli, from which ultramarine has been extracted, varying in colour from dull gray to blue. Although not equal in beauty, and inferior in strength of colour, to ultramarine, they are extremely useful pigments, affording grays much more pure and tender than such as are composed of black and white, or other blues, and better suited to the pearly tints of flesh, foliage, the grays of skies, the shadows of draperies, &c., in which the old masters were wont to employ them. Ultramarine broken with black and white, &c., produces the same effects, and is thus sometimes carried throughout the colouring of a picture.

The brighter sorts of ultramarine ashes are more properly pale ultramarines, and of the class of blue.

IV. PHOSPHATE OF IRON is a native ochre, which classes in colour with the deeper hues of ultramarine ashes, and is eligible for all their uses. It has already been described under its appellation of *blue ochre.*

Slate clays and several native earths class with grays; but the colours of some of the latter, which we have tried, are not durable, being subject to become brown by the oxidation of the iron they contain.

V. PLUMBAGO.   See *Black Lead.*

CHAP. XXI.

OF THE NEUTRAL,

BLACK.

If white and *black* blend, soften, and unite
A thousand ways, is there no black and white?

POPE.

BLACK is the last and lowest in the series or scale of colours descending,— the opposite extreme from white,—the maximum of colour. To be perfect it must be neutral with respect to colours individually, and absolutely transparent, or destitute of reflective power in regard to light; its use in painting being to represent shade or depths, of which it is the element in a picture and in colours, as white is of light.

As there is no perfectly pure and transparent black pigment, black deteriorates all colours in deepening them, as it does warm colours by partially neutralizing them, but it combines less injuriously with cold colours. Though it is the antagonist or contrast of white, yet added to it in minute portion it in general renders white more neutral, solid, and local, with less of the character of light.

As a local colour in a picture, it has the effect of connecting or amassing surrounding objects, and it is the most retiring of colours, which property it communicates to other colours in mixture. It heightens the effect of warm as well as light colours, by a double contrast when opposed to them, and in like manner subdues that of cold and deep colours; but in mixture or glazing these effects are reversed, as we have already said, by reason of the predominance of cold colour in the constitution of black.

Black is to be considered as a synthesis of the three primary colours, the three secondaries, or the three tertiaries, or of all these together—and consequently also of the three semi-neutrals, and may accordingly be composed of due proportions of either tribe or triad. All antagonist colours, or contrasts, also afford the neutral black by composition; but in all the modes of producing black by compounding colours, blue is to be regarded as its archeus or predominating colour, and yellow as subordinate to red, in the proportions, when their hues are true, of eight blue, five red, and three yellow. It is owing to this predominance of blue in the constitution of black, that it contributes by mixture to the pureness of hue in white colours, which in general incline to warmth, and that it produces the cool effect of blueness in glazing and tints, or however otherwise diluted or dilated.

All colours are comprehended in the synthesis of black, consequently the whole sedative power of colour is comprised in black. It is the same in the synthesis of white; and, with like relative consequence, white comprehends all the stimulating powers of colour in painting. It follows that a little black or white are equivalent to much colour, and hence their use as colours requires judgment and caution in painting; and in engraving, black and white supply the place of colours, and hence a true knowledge of the active or sedative power of every colour is of great importance to the engraver, and of main consideration in every mode of the chiar'-oscuro.

By due attention to the synthesis of black it may be rendered a harmonizing medium to all colours, and it gives brilliancy to them all by its sedative effect on the eye, and its powers of contrast; nevertheless, we repeat, as a pigment it must be introduced with caution in painting when *hue* is of greater importance than *shade*, even when employed as shadow. Without great judgment in its use black is apt to appear as local colour, rather than as privation of light; and for this reason deep and transparent colours, which have darkness in their constitution, are better adapted in general for producing the true natural effects of shade. From the contrasting and harmonizing efficacy of black with all colours, and in particular with the lively and gay, the goddess Flora has been not inaptly decorated by mythologists with a mantle of black; and the moral sentiment, arising from the same cause, has not escaped the elegant imagination of the poet, thus beautifully and succinctly expressed by Gray,—

>The *hues of bliss more brightly glow*,
>Chasten'd by *sabler tints of woe*.

Black is emblematical of mental degradation and crime; the garb of the Harpies and Furies, the daughters of *Night*. In its moral effects individually, it is gloomy and terrific both in nature and art;* hence fear and horror are excited and augmented by darkness; hence it has been the livery of woe, and the ensign of death and the devil, among every civilized people; and hence the poets, priests, and rhetoricians have employed it ideally in designating the dismal, the dreadful, the criminal, the mournful, the horrible, and in every sentiment of *melancholy*, of which the very name denotes blackness and darkness. Such also are its expressive uses in painting, in which it is the instrument of solemnity, obscurity, breadth and boundlessness, the terrible, the sublime, and the profound; and it is by contrast the prime power whereby all the magic of the chiar'-oscuro is produced. Upon this power of the neutral, Rembrandt depended as much too much perhaps as Rubens did upon the contrast of positive colours: the works of Rembrandt afford, nevertheless, the best examples of this power generally; and, in the particular department of landscape, we know of none so varied and perfect as the admirable series of mezzotintos recently published by our English Constable, in a mode of engraving, by the bye, peculiarly qualified for exhibiting the powers of the pencil in light and shade.†

If we compare these natural, sensible, and moral powers of black, with those of white, which colours are symbols of night and day, we shall be struck by the immense latitude of light and shade which lies between them, and the correspondence of opposition which belongs to them equally by nature and the consent of mankind, and be led to infer the similar, moral, and sensible analogies of other colours; not as conventional fancies of the poet and painter, but as natural and real relations and attributes, although dimly understood. By insisting too earnestly upon such paradoxical powers, we may, perhaps, incur from some persons the reproach of the ancient musician who deduced every thing from his own art; and yet we confess we are among the most reluctant to join in this censure of the harmonist, being entirely convinced that, *throughout nature and science, the* GREAT AUTHOR OF ALL *does but manifest the same Identical Wisdom in a variety of ways*, and that He has capacitated man to comprehend, to imitate, and to enjoy them.

We have endeavoured to show, under their proper heads, how black is

* The youth couched by Cheselden had an impression of *great uneasiness* the first time he saw black; and was, some months after, *struck with great horror* at the sight of a negress.
† See Landscapes Characteristic of English Scenery, by John Constable, R.A. engraved by D. Lucas.

related to, and how it affects each colour individually in painting. Black, white, red, blue, green, purple, and brown, are the colours of the poet, with whom gold supplies the place of yellow and orange, as it does also in some old paintings and illuminatings; but, a tinge of melancholy being essential to pathos, black is more employed for effect in eloquence and poetry than all the other colours put together; of which, and also of various relations of this colour, the following may serve as illustrations:—

Black is the badge of hell,
The hue of dungeons, and the scowl of night.
SHAKSPEARE.

Not the *black* gates of *Hades* are to me
More hostile or more hateful, than the man
Whose tongue holds no communion with his heart.
SYDENHAM, AFTER HOMER.

Then came this woful Theban, Palemon,
With flotery berd, and ruggy, *ashy* beres,
In clothes *blacke*, ydropped all with teres,
The reufullest of all the compagnie.
CHAUCER'S KNIGHT'S TALE.

News fitted to the night,—
Black, fearful, comfortless, and horrible.
SHAKSPEARE.

The *blacke and doleful ebonie.*
SPENSER'S ELEGY.

Hence, *loathed Melancholy,*
Of Cerberus and *blackest Midnight* born,
In Stygian cave forlorn,
'Mongst horrid shapes, and shrieks, and sights unholy,
Find out some uncouth cell,
Where brooding *Darkness* spreads his jealous wings,
And the *night-raven* sings;
There, under *ebon shades*, and low-brow'd rocks,
As ragged as thy locks,
In dark Cimmerian desert ever dwell.
MILTON, L'ALLEGRO.

'Tis so strange,
That, though the truth of it *stands off as gross*
As white from black, my eyes will scarcely see it.
SHAKSPEARE.

OF THE NEUTRAL, BLACK.

Was I deceived, or did a *sable cloud*
Turn forth her *silver lining on the night?*
I did not err,—there does a *sable cloud*
Turn forth her *silver lining on the night,*
And *casts a gleam over this tufted grove.*
MILTON, COMUS.

Most preposterous event, that draweth
From my *snow-white* pen the *sable-colour'd ink.*
SHAKSPEARE.

Fame, if not double-faced, is double-mouth'd;
And with contrary blasts proclaims most deeds:
On both his wings, one *black* thé other *white,*
Bears greatest names in his wild aery flight.
MILTON.

Black Macbeth will seem as *pure as snow.*
SHAKSPEARE.

Youth no less becomes
The *light and careless livery* that it wears,
Than *settled age his sables and his weeds,*
Importing health and graveness.
IDEM, HAMLET, Act IV. Scene 7.

Stars, hide your *fires!*
Let not *light* see my *black and deep* desires.
IDEM.

'Tis not your *inky brows,* your *black silk hair,*
Your *bugle eye-balls,* nor your *cheek of cream,*
That can entame my spirits to your worship.
IDEM.

The *Devil damn thee black,* thou *cream-faced loon.*
IDEM.

We *mourn in black,* why mourn we not in *blood?*
IDEM.

The *splendid fortune* and the *beauteous face*
    (Themselves confess it, and their Sires bemoan)
Too soon are caught by *scarlet* and by lace;
    The sons of science shine in black alone.
DUNCOMBE.

And sullen Moloch, fled,
Hath left in *shadows dread*
   His *burning idol*, all of *blackest hue;*
In vain with cymbal's ring
They call the *grisly* king,
   In dismal dance about the furnace blue.
MILTON.

   Come, *thick Night,*
And *pall thee in the dunnest smoke of hell;*
That my keen knife see not the wound it makes,
Nor *heaven peep through the blanket of the dark,*
To cry, Hold! Hold!
SHAKSPEARE, MACBETH.

Richard yet lives, *hell's black intelligencer.*
IDEM, RICHARD III.

How now you *secret, black, and midnight hags?*
IDEM, MACBETH.

O, beat away the busy meddling *fiend,*
That lays strong siege unto this wretch's soul,
And from his bosom purge this *black despair!*
IDEM, HENRY VI.

Then comes the *father of the tempest* forth,
Wrapt in black glooms.
THOMSON.

Abhorred Styx, the flood of deadly hate;
Sad *Acheron, of sorrow black and deep.*
MILTON.

O'erlaid with black, staid Wisdom's hue.
IDEM.

   Black perdition.
IDEM.

Besieged with *sable-colour'd Melancholy,*
I did commend the *black oppressing humour*
To the most wholesome physic of thy health-giving air.
SHAKSPEARE.

By heaven! thy love is *black as ebony.*
IDEM.

He said my eyes were *black*, and my hair *black*,
And, now I am remember'd, scorn'd at me.
SHAKSPEARE.

'Tis not alone my *inky cloak*, good mother,
Nor customary suits of *solemn black*,——
Together with all forms, modes, shews of grief,
That can denote me truly——
But I have that within which passeth show.
IDEM, HAMLET, Act I. Sc. 2.

The rugged Pyrrhus,—he, whose *sable arms*,
Black as his purpose, did the night resemble.
IDEM, Act II. Sc. 2.

Arise, black vengeance, from thy hollow cell.
IDEM, OTHELLO, Act III. Sc. 3.

Thus like *the sad presaging raven*, that tolls
The sick man's passport in her hollow beak,
And in *the shadows of the silent night*
Does *shake contagion from her sable wings.*
MARLOWE'S JEW OF MALTA.

When *in dim chambers long black weeds* are seen,
And *weeping's* heard where only *joy* has been.
ROGERS.

Here, while *the proud* their long drawn *pomps display*,
There *the black gibbet glooms* beside the way.
GOLDSMITH.

A cloudy region, *black and desolate*,
Where once a slave withstood a world in arms.
ROGERS.

But the *shadows of eve*, which encompass the *gloom*,
The abode of *the dead*, and the place of *the tomb.*
HERBERT KNOWLES.

With soft suspended step, and *muffled deep
In midnight darkness*, whisper'd my last sigh.
YOUNG.

She hath abated me of half my train;
Look'd black upon me; struck me with her tongue,
Most serpent-like, upon the very heart.
SHAKSPEARE, KING LEAR, Act II. Sc. 4.

> At length they chanced to meet upon the way
> An aged sire, in long *black weeds* yclad,
> His feet all bare, his beard all *hoarie gray*,
> And by his belt his booke he hanging had;
> Sober he seem'd, and very *sagely sad*.
> SPENSER, FAIRIE QUEEN, Canto I. 29.

Most of the black pigments in use are produced by charring, and owe their colour to the carbon they contain: such are *Ivory* and *Bone blacks, Lamp black, Blue black, Frankfort black,* &c. The three first are most in use, and vary according to their modes of preparation or burning; yet fine *Frankfort black*, though principally confined to the use of the engraver and printer, is often preferable to the others.

Native or *mineral blacks* are heavy and opaque, but dry well.

Some of the old masters are said to have employed a *black lake* of great beauty,—and all coloured lakes calcined in close vessels become such; or perhaps they employed the sediment of the dyer's-vat, which Pliny informs us was used by the antients, and which nevertheless could not have been a durable nor eligible pigment, more especially in distemper or fresco. It is probable also that this black lake may have been a synthetic black, composed of primary or secondary transparent colours, or by addition of coloured lakes to other blacks as the case might require. Prussian blue and burnt lake afford a powerful black; and *compound blacks,* in which transparent pigments are employed, will generally go deeper and harmonize better with other colours than any original black pigment alone: hence lakes and deep blues, added to the common blacks, greatly increase their clearness and intensity; and ultramarine has evidently been employed in mixture and glazing of the fine blacks of some old pictures. In this view, black altogether compounded of ultramarine, cobalt blue, or Prussian blue, with red and yellow lakes, subordinated according to the powers of the pigments used, will afford the most powerful and transparent blacks; but they dry badly in oil, as is indeed the case also of most other blacks.

Black pigments are innumerable: the following are however the principal, all of which are permanent colours:—

I. IVORY BLACK and *Bone Black* are ivory and bone charred to blackness by strong heat in closed vessels. These pigments vary principally through want of care or skill in preparing them: when well made, they are fine neutral blacks, perfectly durable and eligible both for oil and water painting; but when insufficiently burnt they are brown, and dry badly; and when too much burnt, they are cineritious, opaque, and faint in colour. Of the two, ivory affords the best pigment; but bone black is commonly used, and immense quantities are consumed with sulphuric acid in manufacturing of *shoe-blacking*.

II. LAMP-BLACK, or *Lamblack*, is a smoke black, being the soot of resinous woods, obtained in the manufacturing of tar and turpentine. It is a pure carbonaceous substance of a fine texture, intensely black, and perfectly durable, but dries badly in oil. This pigment may be prepared extemporaneously for water painting by holding a plate over the flame of a lamp or candle, and adding gum-water to the colour: the nearer the plate is held to the wick of the lamp, the more abundant and warm will be the hue of the black obtained; at a greater distance it will be more effectually charred and blacker. This is a good substitute for Indian ink, the colouring basis of which appears to be lamp-black.

III. FRANKFORT BLACK is said to be made of the lees of wine from which the tartar has been washed, by burning, in the manner of ivory black. Similar blacks are prepared of *vine twigs and tendrils*, which contain tartar; also from *peach-stones*, &c. whence *Almond black*; and the Indians employ for the same purpose the *shell of the cocoa-nut*: and inferior Frankfort black is merely the levigated charcoal of woods, of which the hardest, such as the *box* and *ebony*, afford the best. Fine Frankfort black, though almost confined to copper-plate printing, is one of the best black pigments we possess, being of a fine neutral colour, next in intensity to lamp black, and more powerful than that of ivory. Strong light has the effect of deepening its colour; yet the blacks employed in the printing of engravings have proved of very variable durability.

IV. BLUE BLACK is also a well-burnt and levigated charcoal, of a cool neutral colour, and not differing in other respects from the common Frankfort black above mentioned. Blue black was formerly much employed in painting, and, in common with all carbonaceous blacks, has, when duly mixed with white, a preserving influence upon that colour in two respects, which it owes, chemically, to the bleaching power of carbon, and, chromatically, to the neutralizing and contrasting power of black with white. It would be well also for the art if carbon had a like power upon the colour of oils; but of this it is deficient; and although chlorine destroys their colour temporarily, they re-acquire it at no very distant period.

V. SPANISH BLACK is a soft black, prepared by burning *cork* in the manner of Frankfort and ivory blacks; and it differs not essentially from the former, except in being of a lighter and softer texture. It is subject to the variation of the above charred blacks, and eligible for the same uses.

VI. PURPLE BLACK is a preparation of madder of a deep purple hue approaching black: its tints with white lead are of a purple colour. It is very transparent and powerful, glazes and dries well in oil, and is a durable and eligible pigment, more properly belonging perhaps to the semi-neutral class of marrone.

VII. MINERAL BLACK is a native impure oxide of carbon, of a soft texture, found in Devonshire. It is blacker than plumbago, and free from its metallic lustre,—is of a neutral colour, greyer and more opaque than ivory black,—forms pure neutral tints,—and being perfectly durable, and drying well in oil, it is valuable in dead colouring on account of its solid body, as a preparation for black and deep colours before glazing. It would also be the most durable and best possible black for frescos.

VIII. BLACK OCHRE is a variety of the above, combined with iron and alluvial clay. It is found in most countries, and should be washed and exposed to the atmosphere before it is used. Sea-coal, and innumerable black mineral substances have been and may be employed as succedanea for the more perfect blacks, when the latter are not procurable, which rarely happens.

IX. BLACK CHALK is an indurated black clay, of the texture of white chalk, and is naturally allied to the two preceding articles. Its principal use is for cutting into the crayons, which are employed in sketching and drawing.

Fine specimens have been found near Bantry in Ireland, and in Wales, but the Italian has the best reputation. Crayons for these uses are also prepared artificially, which are deeper in colour and free from grit. Charcoal of wood is also cut into crayons for the same purpose.

X. INDIAN INK. The pigment well known under this name is principally brought to us from China in oblong cakes, of a musky scent, ready prepared for painting in water; in which use it is so well known, and so generally employed, as hardly to require naming. It varies, however, considerably in colour and quality, and is sometimes, properly, called *China ink*. Various accounts are given by authors of the mode of preparing this pigment, the principal substance or colouring matter of which is a smoke black, having all the properties of our lamp-black, and the variety of its hues and texture seems wholly to depend upon the degree of burning and levigating it receives.

XI. BLACK LEAD, *Plumbago*, or *Graphite*, is a native carburet of iron or oxide of carbon, found in many countries, but nowhere more abundantly nor so fine in quality as at Borrodale in Cumberland, where there are mines of it, from which the best is obtained, and consumed in large quantity in the formation of crayons and the black-lead pencils of the shops, which are in universal use in writing, sketching, designing, and drawing—for which the facility with which it may be rubbed out by Indian rubber, caoutchouc, or gum-elastic, and the crumb of bread, admirably adapts it.*

Although not acknowledged as a pigment, its powers in this respect claim a place for it, at least among water-colours; in which way, levigated in gum-water in the ordinary manner, it may be used effectually with rapidity and freedom in the shading and finishing of pencil drawings, &c., and as a substitute therein for Indian ink. Even in oil it may be useful occasionally, as it possesses remarkably the property of covering, forms *grey* tints, dries quickly, injures no colour chemically, and endures for ever.

* Drawings, &c. in pencil are sometimes required to be fixed. This is best and most easily done with *water-starch*, prepared in the manner of the laundress, of such strength as just to form a jelly when cold, which may be then applied with a broad camel's-hair brush, as in varnishing. The same may be done with thin, cold isinglass size, or rice-water.

CHAP. XXII.

TABLES OF PIGMENTS.

As there are circumstances under which some pigments may very properly and safely be used, which under others might prove injurious or destructive to the work, the following Lists or Tables are subjoined, in which they are classed according to various general properties, as guides to a judicious selection. These Tables are the results of direct experiments and observation, and are composed, without regard to the common reputation or variable character of pigments, according to the real merits of the various specimens tried.

The powers of pigments therein adverted to might have been denoted by numbers; but since there is no exact and constant agreement in different specimens of like pigments, nor relatively among different pigments, it would have been an affectation of accuracy without utility: add to which, the properties and effects of pigments are much influenced by adventitious circumstances, and are sometimes varied or altogether changed by the grounds on which pigments are used; by the vehicles in which they are used; by the siccatives and colours with which they are used; and by the varnishes by which they are covered.

These Tables are therefore offered only as approximations to the true characters of pigments, (some of which, for the above reasons, are liable to be disputed,) and as general guides to right practice. They render it also apparent as a general conclusion, that the majority of pigments have a mediocrity of qualification balancing their excellences with their defects, and that the number of good and eligible pigments overbalances those which ought in general to be rejected.

TABLE I.

Of Pigments, the colours of which suffer different degrees of change by the action of light, oxygen, and pure air; but are little, or not at all, affected by shade, sulphuretted hydrogen, damp, and foul air:—

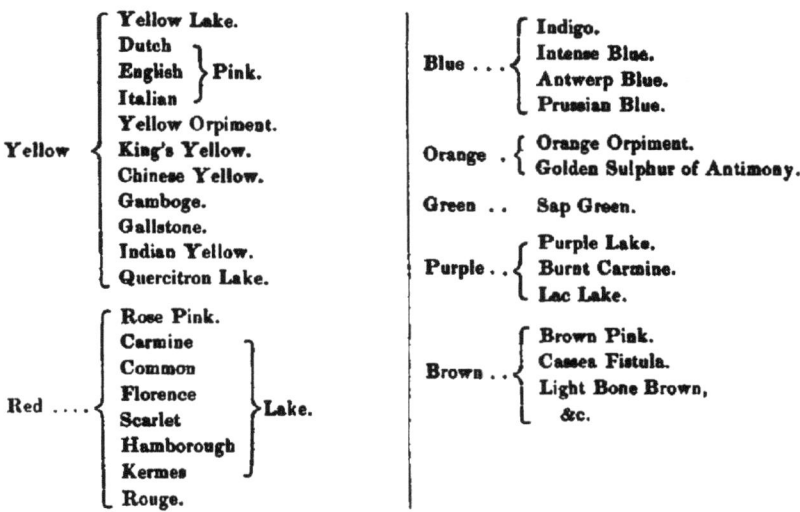

Yellow { Yellow Lake. Dutch, English, Italian } Pink. Yellow Orpiment. King's Yellow. Chinese Yellow. Gamboge. Gallstone. Indian Yellow. Quercitron Lake.

Red .... { Rose Pink. Carmine, Common, Florence, Scarlet, Hamborough, Kermes } Lake. Rouge.

Blue ... { Indigo. Intense Blue. Antwerp Blue. Prussian Blue.

Orange . { Orange Orpiment. Golden Sulphur of Antimony.

Green .. Sap Green.

Purple .. { Purple Lake. Burnt Carmine. Lac Lake.

Brown .. { Brown Pink. Cassea Fistula. Light Bone Brown, &c.

REMARKS.—None of the pigments in this Table are eminent for permanence. No white or black pigment whatever belongs to this class, nor does any tertiary, and a few only of the original semi-neutrals. Most of those included in the list fade or become lighter by time, and also, in general, less bright.

TABLE II.

PIGMENTS, the colours of which are little, or not at all, changed by light, oxygen, and pure air; but are more or less injured by the action of shade, sulphuretted hydrogen, damp, and impure air:—

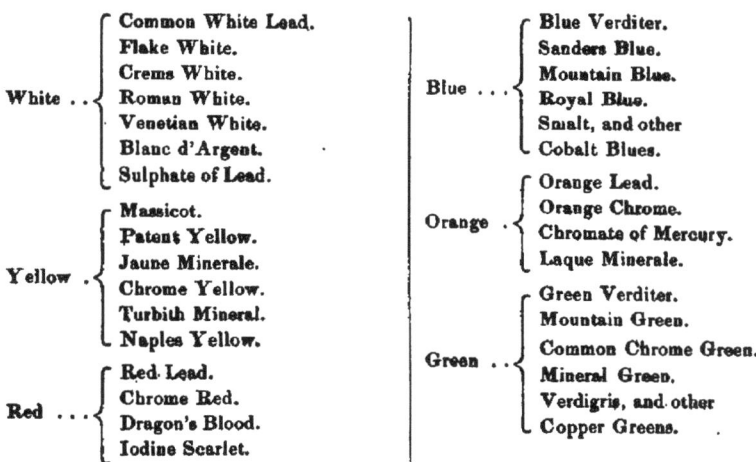

White ..
- Common White Lead.
- Flake White.
- Crems White.
- Roman White.
- Venetian White.
- Blanc d'Argent.
- Sulphate of Lead.

Yellow .
- Massicot.
- Patent Yellow.
- Jaune Minerale.
- Chrome Yellow.
- Turbith Mineral.
- Naples Yellow.

Red ...
- Red Lead.
- Chrome Red.
- Dragon's Blood.
- Iodine Scarlet.

Blue ...
- Blue Verditer.
- Sanders Blue.
- Mountain Blue.
- Royal Blue.
- Smalt, and other
- Cobalt Blues.

Orange .
- Orange Lead.
- Orange Chrome.
- Chromate of Mercury.
- Laque Minerale.

Green ..
- Green Verditer.
- Mountain Green.
- Common Chrome Green.
- Mineral Green.
- Verdigris, and other
- Copper Greens.

REMARKS.—Most of our best white pigments are comprehended in this Table, but no black, tertiary, or semi-neutral colour.

Many of these colours, when secured by oils and varnish, &c., may be long protected from change. The pigments of this Table may be considered as more durable than those of the preceding; they are nevertheless ineligible in a water-vehicle, particularly for miniature painting, and most of them become darker by time alone in every mode of use.

This list is the opposite of Table I.

TABLE III.

Pigments, the colours of which are subject to change by the action both of light and oxygen, and the opposite powers of sulphuretted hydrogen, damp, and impure air:—

White ..	{ Pearl, or Bismuth White. Antimony White. Submuriate of Mercury.		Blue ...	{ Royal Blue. Prussian Blue. Antwerp Blue.
Yellow .	{ Turbith Mineral. Patent Yellow.		Orange..	{ Sulphate of Antimony. Anotta. Carucru.
Red....	{ Iodine Scarlet. Dragon's Blood.		Green .. Russet ..	Verdigris. Prussiate of Copper.

Remarks.—This Table comprehends our most imperfect pigments, and demonstrates how few absolutely bad have obtained currency. Indeed several of them are valuable for some uses, and not subject to sudden or extreme change by the agencies to which they are here attributed. Yet the greater part of them are destroyed by time.

These pigments unite the bad properties of those in the two preceding Tables.

TABLE IV.

PIGMENTS, not at all or little liable to change by the action of light, oxygen, and pure air, nor by the opposite influences of shade, sulphuretted hydrogen, damp and impure air; nor by the action of lead or iron:—

White ..
- Zinc White.
- True Pearl White.
- Constant or Barytic White.
- Tin White.
- The Pure Earths.

Yellow .
- Yellow Ochre.
- Oxford Ochre.
- Roman Ochre.
- Sienna Earth.
- Stone Ochre.
- Brown Ochre.
- Platina Yellow.
- Lemon Yellow.

Red ....
- Vermilion.
- Rubiates, or Madder Lakes.
- Madder Carmines.
- Red Ochre.
- Light Red.
- Venetian Red.
- Indian Red.

Blue ...
- Ultramarine.
- Blue Ochre.

Orange ..
- Orange Vermilion.
- Orange Ochre.
- Jaune de Mars.
- Burnt Sienna Earth.
- Burnt Roman Ochre.
- Damonico.
- Light Red, &c.

Green ...
- Chrome Greens.
- Terre-Verte.
- Cobalt Green.

Purple ..
- Gold Purple.
- Madder Purple.
- Purple Ochre.

Russet ..
- Russet Rubiate, or Madder Brown.
- Intense Russet.
- Orange Russet.

Brown and Semi-neutral.
- Vandyke Brown.
- Rubens' Brown.
- Bistre.
- Raw Umbre.
- Burnt Umbre.
- Marrone Lake.
- Cassel Earth.
- Cologne Earth.
- Antwerp Brown.
- Hypocastanum, or Chestnut Brown.
- Asphaltum.
- Mummy, &c.
- Phosphate of Iron.
- Ultramarine Ashes.
- Sepia.
- Manganese Brown.

Black ..
- Ivory Black.
- Lamp Black.
- Frankfort Black.
- Mineral Black.
- Black Chalk,
- Indian Ink.
- Graphite.

REMARKS.—This Table comprehends all the best and most permanent pigments, and such as are eligible for water and oil painting. It demonstrates that the best pigments are also the most numerous, and in these respects stands opposed to the three Tables preceding.

TABLE V.

PIGMENTS subject to change variously by the action of white lead and other pigments, and preparations of that metal.

Yellow .	Massicot. Yellow Orpiment. King's Yellow. Chinese Yellow. Gamboge. Gall-stone. Indian Yellow. Yellow Lake. Dutch ⎫ English ⎬ Pink. Italian ⎭		Blue	Indigo.
			Orange	⎧ Orange Lead. ⎨ Orange Orpiment. ⎨ Golden Sulph. of Antimony. ⎨ Anotta, or Roucou. ⎩ Carucru, or Chica.
			Green ..	Sap Green.
			Purple ..	⎧ Purple Lake. ⎩ Burnt Carmine.
Red ...	Iodine Scarlet. Red Lead. Dragon's Blood. Common ⎫ Cochineal ⎪ Florence ⎬ Lake. Scarlet ⎪ Hambro' ⎪ Lac ⎭ Carmine. Rose Pink.		Citrine ..	⎧ Brown Pink. ⎩ Cassia Fistula.

REMARKS.—Acetate or sugar of lead, litharge and oils, rendered drying by oxides of lead, are all in some measure destructive of these colours. Light, bright, and tender colours are principally susceptible of change by the action of lead.

The colours of this Table are very various in their modes of change, and thence do not harmonize well by time: it follows too that when any of these pigments are employed, they should be used pure or unmixed; and, by preference, in varnish, while their tints with white lead ought to be altogether rejected.

TABLE VI.

PIGMENTS, the colours of which are subject to change by iron, its pigments, and other ferruginous substances.

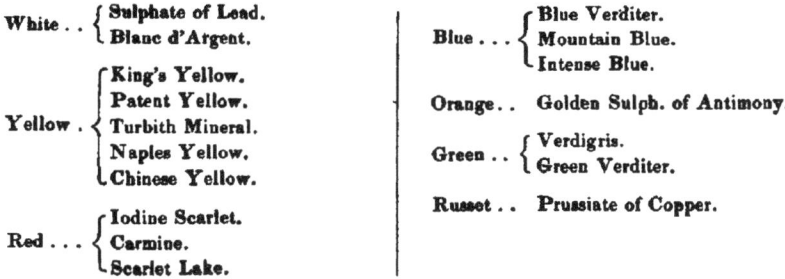

White .. { Sulphate of Lead. / Blanc d'Argent.

Yellow . { King's Yellow. / Patent Yellow. / Turbith Mineral. / Naples Yellow. / Chinese Yellow.

Red ... { Iodine Scarlet. / Carmine. / Scarlet Lake.

Blue ... { Blue Verditer. / Mountain Blue. / Intense Blue.

Orange.. Golden Sulph. of Antimony.

Green .. { Verdigris. / Green Verditer.

Russet.. Prussiate of Copper.

REMARKS.—Several other delicate pigments are slightly affected by iron and its preparations; and with all such, as also with those of the preceding Table, and with all pigments not well freed from acids or salts, the iron palette-knife is to be avoided or used with caution, and one of ivory or horn substituted in its place. Nor can the pigments of this Table be in general safely combined with the ochres.

Strictly speaking, that degree of friction which abrades the palette-knife in rubbing of colours therewith is injurious to every bright colour.

TABLE VII.

PIGMENTS more or less transparent, and generally fit to be employed as glazing and finishing colours, if not disqualified according to Tables I, II, III.

Yellow .	Platina Yellow. Sienna Earth. Gamboge. Indian Yellow. Gall-stone. Quercitron Lake. Italian ⎫ English ⎬ Pink. Dutch ⎭ Yellow Lake.	Purple .	Madder Purple. Burnt Carmine. Purple Lake. Lac Lake.
		Citrine .	Brown Pink. Citrine Lake. Casia Fistula.
		Russet .	Madder Brown. Prussiate of Copper.
		Olive . . .	Olive Lake.
Red . . .	Madder Carmine. Madder Lakes. Lac Lake. Carmine. Kermes ⎫ Common ⎪ Florence ⎬ Lakes. Scarlet ⎪ Hambro' ⎭ Dragon's Blood. Rose Pink.	Brown .	Vandyke Brown. Cologne Earth. Burnt Umbre. Bone Brown. Asphaltum. Mummy. Brown Pink. Antwerp Brown. Bistre. Sepia. Chestnut Brown. Prussian Brown.
Blue . . .	Ultramarines. Cobalt Blue. Smalt. Royal Blue. Prussian Blue. Antwerp Blue. Intense Blue. Indigo. Blue Ochre.	Marrone .	Marrone Lake. Carucru.
		Gray . . .	Ultramarine Ashes. Phosphate of Iron.
		Black . .	Ivory Black. Bone Black. Lamp Black. Frankfort Black. Blue Black. Spanish Black. Purple Black.
Orange .	Madder Orange. Anotta. Burnt Sienna Earth. Jaune de Mars.		
Green . .	Chrome Green. Sap Green. Prussian Green. Terre-Verte. Verdigris.		

REMARKS.—This Table comprehends most of the best water-colours.

Pigments not inserted in this Table may of course be considered of an opposite class, or *opaque* colours; with which, nevertheless, transparent effects in painting are produced by the skill of the artist in breaking and mingling without mixing them, &c.

TABLE VIII.

PIGMENTS, the colours of which are little or not at all affected by heat or fire.

White ....
- Tin White.
- Barytic White.
- Zinc White.
- The Pure Earths.

Yellow ..
- Naples Yellow.
- Patent Yellow.
- Antimony Yellow.

Red ....
- Red Ochre.
- Light Red.
- Venetian Red.
- Indian Red.

Blue ....
- Royal Blue.
- Smalt.
- Dumont's Blue, and all Cobalt Blues.
- Ultramarines.

Orange ...
- Orange Ochre.
- Jaune de Mars.
- Burnt Sienna Earth.
- Burnt Roman Ochre.
- Damonico.

Green ...
- Chrome Green.
- Cobalt Green.

Purple ...
- Gold Purple.
- Purple Ochre.

Brown ...
- Rubens' Brown.
- Burnt Umbre.
- Cassel Earth.
- Cologne Earth.
- Antwerp Brown.
- Manganese Brown.

Black ...
- Graphite.
- Mineral Black.

REMARKS.—Many of the pigments of this Table are available in enamel-painting, and most of them are durable in the other modes.

TABLE IX.

PIGMENTS which are little or not at all affected by *lime*, and in various degrees eligible for fresco, distemper, and crayon painting.

White ...	Barytic White. Pearl White. Gypsum, and all Pure Earths.	Green ...	Green Verditer. Mountain Green. Chrome Green. Mineral Green. Emerald Green. Verdigris, and other Copper Greens. Terre-Verte. Cobalt Green.
Yellow ..	Yellow Ochre. Oxford Ochre. Roman Ochre. Sienna Earth. Di Palito. Stone Ochre. Brown Ochre. Indian Yellow. Patent Yellow. Naples Yellow. Massicot.	Purple ...	Gold Purple. Madder Purple. Purple Ochre.
Red	Vermilion. Red Lead. Red Ochre. Light Red. Venetian Red. Indian Red. Madder Reds.	Brown and Semineutral	Bone Brown. Vandyke Brown. Rubens' Brown. Bistre. Raw Umbre. Burnt Umbre. Cassel Earth. Cologne Earth. Antwerp Brown. Chestnut Brown. Asphaltum. Mummy. Ultramarine Ashes. Manganese Brown.
Blue	Ultramarine. Smalt, and all Cobalt Blues.		
Orange ..	Orange Vermilion. Orange Lead. Orange Chrome. Laque Mineral. Orange Ochre. Jaune de Mars. Burnt Sienna Earth. Damonico. Light Red, &c.	Black ...	Ivory Black. Lamp Black. Frankfort Black. Mineral Black. Black Chalk. Indian Ink. Graphite.

REMARKS.—This Table shows the multitude of pigments from which the painters in fresco, distemper, and crayons, may select their colours; in doing which, however, it will be necessary they should consult the previous Tables respecting other qualities of pigments essential to their peculiar modes of painting; and as these modes are exciting renewed interest in the world of art, tending to their extension in practice, particularly the latter of them, we will subjoin a few remarks.

ON FRESCO, &c.

The art of painting in fresco is so naturally adapted to the grandeur of historical and patriotic painting, to which it appears to have been first applied, and the zealous attention of eminent artists being at present turned to the revival of this great and free mode of art, we will not withhold the few observations we have made thereon in connexion with colours and colouring, however brief our experience may have been.

It is hardly necessary to inform the reader, that *fresco painting* is performed with pigments prepared in water, and applied upon the surface of *fresh laid plaster* of lime and sand, with which walls are covered; and as it is that mode of painting which is least removed in practice from modelling or sculpture, it might not improperly be called *plastic painting*.

As lime, in an active state, is the common cementing material of the ground and colours employed in fresco, it is obvious that such colours or pigments only can be used therein as remain unchanged by lime. This need not, however, be a universal rule for painting in fresco, since other cementing materials, as strong or stronger than lime, may be employed, which have not the action of lime upon colours—such is calcined gypsum, of which plaster of Paris is a species; which, being neutral sulphats of lime, exceedingly unchangeable, have little or no chemical action upon colours, and would admit even Prussian blue, vegetal lakes, and the most tender colours to be employed thereon, so as greatly to extend the sphere of colouring in fresco, adapted to its various design; which bases merit also the attention of the painter in crayons and distemper.

So far too as regards durability and strength of the ground, the Parker's cement, now so generally employed in architectural modellings, would afford a new and advantageous ground for painting in fresco; and, as it resists damp and moisture, it is well adapted, with colours properly chosen, to situations in which paintings, executed in other modes of the art, or even in ordinary fresco, would not long endure.

As these materials, and others now in use, were either unknown or unemployed by the antient painters in fresco, their practice was necessarily limited to the pigments enumerated in the preceding table; but every art demands such a variation in practice, as adapts it to circumstances and the age in which it is exercised, without attention to which it may degenerate, or, at best, remain stationary, but cannot advance.

In point of durability, however, both as respects colours and texture, the frescos of the antient Egyptians (if they may be so called) have alone pretensions to the character of almost perpetual incorruptibility; in which respect fresco must have declined, at the same time that painting as an art advanced, even among the Greeks; while many of the earlier works of the moderns, founded on the basis of Grecian art, have nothing to boast of in this respect; and the "Last Judgment" of Michael Angelo, and many great performances, may be adduced as examples thereof; so that, aided by modern physics, we may hope not only the restoration, but improvement of this art.

Although differing exceedingly in their mechanical execution, the modes of fresco, distemper, and crayon painting agree in their chemical relations; so far, therefore, as respects colours and pigments, the foregoing remarks apply to these latter arts. In distemper painting, however, the carbonate of lime, or whitening employed as a basis, is less active than the pure lime of fresco. The vehicles of both modes are the same, and their practice is often combined in the same work: water is their common vehicle; and to give adhesion to the tints and colours in distemper painting, and make them keep their place, they are variously mixed with the size of glue (prepared commonly by dissolving about four ounces of glue in a gallon of water). Too much of the glue disposes the design to crack and peel from the ground; while, with too little, it is friable and deficient of strength. In some cases the glue may be abated, or altogether dispensed with, by employing plaster of Paris diluted and worked into the colours; by which they will acquire the consistency and appearance of oil paints, without destroying their limpidness, or allowing the colours to separate, while they will acquire a good surface, and keep their place in painting with sufficient strength, and without being liable to mildew,—to which animal glue is disposed, and to which milk, and other vehicles recommended in this mode, are also subject.

Of more difficult introduction in these modes of painting is *bees'-wax*, although it has been employed successfully in each of them, in the encaustic of the antients, &c.; the body colours of the moderns; and, with excellent effect, in crayons,—first, we believe, by the late Mr. Adam Buck. Wax is a most incorruptible substance, and communicates many of the qualities of oil-painting.

CHAP. XXIII.

ON VEHICLES AND VARNISHES.

> How many fondly waste the studious hour
> To seek in process what they want in power,
> Till all in gums engross'd, macgilps, and oils,
> The painter sinks amid the chemist's toils.
>
> SHEE.

SINCE colours and pigments are liable to material influence, and changes of effect, from the liquids employed in painting for tempering, combining, distributing, and securing them on their grounds in the various modes of the art,—the powers and properties of vehicles and varnishes are of hardly less importance than those of colours themselves; they are therefore an essential branch of our subject, and an inquiry of interminable interest among artists. Vehicles are, indeed, among the chief materials and indispensable means of painting, and give names to its principal modes or genera, under the titles of painting in water, in oil, in varnish, &c.: we will consider them in reference to each.

Though originally few and simple, vehicles have been extremely diversified by composition and addition, suited to the various purposes and fancies of artists, so as to have become a subject of no mean extent and intricacy; to explicate which perfectly is as far from our hope as our intention, which is to treat of it in a general way, with such hints and remarks as have sprung from our own observation and experience, and may tend to improvement in practice.

It is observable that the colours of pigments bear out with effects differing according to the liquids with which they are combined, and the substances those liquids hold in solution, which in some instances obscure or depress, and in others enliven or exalt the colours; in the first case by the tinge and opacity of the fluid; and in the latter, by its colourless transparency, and sometimes also much more so by a refractive power,—as in varnishes made of pure resinous substances, which have a very evident and peculiarly exalting effect upon colours, that continues when they are dry; because resins form a glossy transparent cement, while the media, formed by expressed oils, become horny, or semi-opaque. And this principle applies also to aqueous and spirituous vehicles in water-painting, according to the nature of the gums, or other substances they may hold in solution.

As the action of AQUEOUS LIQUIDS, and their solvends upon colours, is stronger and more immediate than that of oils and varnishes, it is of great importance to the water-colour painter that he should attend to the pureness of his water. He ought to use no other than *distilled water;* or, wanting this, he should use *rain-water filtered,* which is next in purity to distilled water. In all hard and impure waters, colours are disposed to separate and curdle, so that it is often impossible a clear flowing wash, or gradation of colour, should be obtained with them. Solution of gums, oxgall, &c. correct, without entirely overcoming these defects of the water; but they are often inconvenient, if not injurious: we recommend therefore to colourmen to keep distilled water ready for artists' use; or the latter may, if he pleases, procure it of the chemists, or use in its stead the distilled water of roses, or lavender, &c. which have no injurious effect upon colours, and recommend themselves by their agreeable scents; but then they must be the really distilled waters, and not the compounds sold as perfumes under their names.

GUM is a necessary addition to water to give pigments their requisite cohesion, and to attach the colours to the paper or ground on which they are applied, as well as to give them the property of bearing out to the eye, according to the intention of the artist; upon which, and upon the pigments used, depend the proportions of gum to be employed, gum being a constituent of some pigments, while others are of textures to require it in considerable quantity to give them proper tenacity,—qualities we have adverted to in speaking of individual pigments: as a general rule, however, the proportion of gum employed with a colour should be sufficient to prevent its abrasion, but not so much as to occasion its scaling or cracking, both of which are easily determined by trial upon paper.

Of GUMS,—SENEGAL is the strongest and best suited to dark colours; but

GUM ARABIC is in general clearer and whiter, and thence is better adapted to the brighter and more delicate colours; these should be purified by solution in water, straining, and decanting, and should be used fresh or preserved by addition of alcohol.

TRAGACANTH is a strong colourless gum, soluble in hot water, and of excellent use when colours are required to lie flat, or not bear out with gloss, and also when a gelatinous texture of the vehicle is of use to preserve the touch of the pencil and prevent the flowing of some colours; for which purpose also solution of *isinglass* is available, and of greater power.

AMMONIAC, or *Gum Ammoniac*, is a gum-resin, soluble in spirit and in water, in the latter of which it forms a milky fluid that dries transparent: it has many properties which render it useful in water-painting, and is, we have found, superior to the gums in forming some colours into cakes, causing them to work off. It is avoided by insects, is very tenaceous, and affords a middle vehicle between oil and water, with some of the advantages of both.

A most excellent *mucilage* for water-painting may be made by diluting gradually clear size of ISINGLASS with boiling water, till, on becoming cold, it just flows and loses its gelatinous texture : in this liquid is then to be dissolved by gentle heat as much colourless gum Senegal, or Arabic, as it will conveniently take up.

The ingenious Mr. Robertson, of Worton, has employed isinglass in water-painting with the happiest effect, of which his well-known admirable copy of the "Bacchus and Ariadne" of Titian is sufficient evidence. This picture, as well as the original works of this gentleman, possess the full powers of oil-painting, with a permanence of tone not to be expected in oil; and being varnished with white lac varnish, may be considered as of extreme durability. His vehicle, for which the Society of Arts voted him a gold medal, may be prepared by suffering shreds of isinglass to imbibe cold water till thoroughly soft, and then dissolving them in boiling alcohol, in such proportions as will just produce a fluid compound when cold. Spirit of wine alone will, Mr. Robertson states, by boiling dissolve isinglass, which we presume is effected by the weakening of the spirit, as the more volatile portion becomes dissipated by the heat.

It has been an opinion of eminent judges in the art, that the Venetian painters—after a mode also ascribed to some of the great antient Greek artists—employed oils and varnishes only as preservatives and defences of their works, and not as vehicles in painting them, for which latter purpose they are supposed to have used water with proper additions; and it has been proposed that artists should adopt the Indian process of painting, in which lac is rendered saponaceous and miscible in water by the medium of borax; but against this process the foul colour and opacity of the vehicle has been justly objected. If, however, one part of borax be dissolved in twelve of boiling water, and the solution be added in equal, or other proportions, to white lac varnish, a perfectly transparent colourless liquid is formed, which diffuses freely in water, and may be used, with some difficulty, as a vehicle for painting instead of oil. Pictures wrought in such vehicle would dry readily, and, being varnished with the white lac varnish, would have a homogeneity of texture throughout, freshness of colouring, and permanence in every way equal to oil-painting; add to this, that as this lac vehicle is as freely miscible with oil as it is with water, it supplies an intermediate mode of painting, or connecting link between painting in water and oil, which probably may, in some ingenious hand, be free from the evils and unite the advantages of both: yet, as the tenacity and adhesion of lac vehicles depend in working upon a higher temperature than is common to our climate, they will in general require artificial heat of the painting-room, except during the height of summer, and are hence better suited to a climate like that of India. The amiable and accomplished Mr. W. H. Watts has been the first to encounter the difficulties of this process.

By similar means, mastic and other soft resins may be rendered miscible in water, and even oils and bees'-wax may be introduced therein. These have the disadvantage, however, of being opaque, although in drying they become transparent.

The mode of encaustic painting, invented by Miss Greenland (afterwards Mrs. Hooker), published in the Transactions of the Society of Arts, &c. for 1792 and 1807, which was an improvement upon the method of Count Caylus, was performed with a vehicle of this kind, which may be prepared by dissolving four and a half ounces of gum Arabic in eight ounces of pure water in a glazed vessel, to which seven ounces of powdered mastic are to be added, and the whole then stirred over a moderate fire till combined in an opaque uniform paste; five ounces of white wax are then to be added, and stirred till melted and beginning to boil, when the vessel should be removed from the fire, and sixteen ounces more of pure cold water gradually stirred into the mixture, which then will form a cream-like composition, to be kept in a bottle for use.

The Society of Arts, &c., have also recently presented a medal to Mr. J. Hammond Jones, for a process in miniature painting, in which he employed as a vehicle a cold saturated solution of borax in water, in one quart of which he dissolved a quarter of an ounce of gum tragacanth, which dried sufficiently firm to allow tints to be repeatedly laid one over another without moving or washing up.

Albumen, or white of egg, has been employed as an addition to water vehicles; and for the common purposes of distemper painting, *milk* and the *serum of blood* have been proposed, but are not adapted to fine art.

The unnecessary use of *sugar* in water-colours should be avoided, as it is disposed to acid fermentation with gum, and is attractive of damp from the atmosphere, and of flies and other insects. *Animal gall* is necessary only to attach the colours to the ground when it rejects them, or they work greasy, as is often the case on ivory and very smooth vellum or polished substances, or over certain pigments. *Borax*, which is mildly alkaline, answers the same purpose. Spirit of wine, or *alcohol*, is principally of use in water vehicles, as an antiseptic, to preserve them from frost, mildew, and putrescence.

Water, as a vehicle compared with *oil*, is of simple and easy use, drying readily, and being subject to little alteration of colour or effect subsequently; for, notwithstanding oils and varnishes are less chemically active upon colours than aqueous fluids are, those vehicles of the oil-painter subject him to all the perplexities of their bad drying, change of colour, blooming, and cracking,—to habits varying with a variety of pigments, and to the contrariety of qualities, by which they are required to unite tenuity with strength, and to be fluid without flowing, &c.; to provide for and reconcile all which has continually exercised the ingenuity of the artist.

The early painters in oil appear to have proceeded in the manner of water-colour drawing;—beginning to sketch in on a white ground, and producing their effects with transparent colours—embossing their lights with opaque tints and body colours; hence much of their freshness, and hence it has been imagined, not without apparent reason, that they commenced their works or sketched in their designs in water-colours. This is evidently contrary to the general mode of modern practice among the best colourists from the time of the Venetians, who are supposed sometimes to have employed the above method, but generally commenced upon coloured grounds, with opaque colours, and finished by glazing with such as are transparent. In the prevailing method of artists there is, however, occasional combination of both the above methods, by alternately painting and glazing throughout the progress of their works.

In the infancy, and for some time after the invention of oil-painting,* *expressed oils* of a drying quality appear to have been used in a simple state, or with the mere addition of substances to assist their drying, or the sole preparation of boiling or fattening. Such vehicles would undoubtedly give to pictures great strength and durability; but it unfortunately happens, that precisely in proportion to the natural strength and power of drying in oils, is their propensity to acquire colour and become dark by age and seclusion from light; hence one principal cause of the various changes, additions, and compositions of these vehicles, by which the stronger and more drying oils of linseed, &c., have in the practice of the artist sometimes given place to the paler and weaker *oils of nuts and poppy*, which dry with more difficulty, yet ultimately also acquire murky colour in proportion to the body in which they are applied, but of a less offensive hue than that of linseed. Even *olive oil*, which is almost wholly destitute of drying power, but is not subject to acquire colour or lose its transparency, is said to have been substituted in the climate of Italy in place of the desiccative oils, but was more probably resorted to as a diluent, like the volatile *essential oils of turpentine, lavender*, &c., which, though destitute of strength, dry rapidly and do not change colour, and by attenuating the drying oils preserve in some measure their

* We speak here of the *re-invention* or improvement of painting in oil by Van Eyck or John of Bruges, about the beginning of the fourteenth century, for it is hardly to be supposed that this process was not much more antient, and had merely fallen into disuse. Vitruvius indeed asserts, that it was employed by the Greeks and Romans in works exposed to weather; and we have it indeed upon record, that in the early time of Grecian art Protogenes had been bred a shippainter; the very existence of which art almost proves the use of oil or varnish therein, capable of protecting colours from the action of water, which wax alone could hardly accomplish. That flax was known to the antient Egyptians, and linseed used by them as food, we have the testimony of Plutarch in his Treatise of Isis and Osiris—we infer, therefore, that its oil must have been known to the Greeks and Egyptians, who could not in such case but have been acquainted with its drying property in climates like theirs. We are indeed too apt to imagine those things to be of recent invention, of which we have neither tradition nor history.

With regard to later times, Walpole and others have related proofs of the existence of oil-painting in Britain long previously to Van Eyck; and the question has been widely investigated by Raspe, in a treatise " On the Discovery of Oil Painting."—4to. 1781.

colour in painting, in proportion as they lower their strength. Of these essential oils, that commonly called spirit of turpentine, employed by painters, is a very useful addition to those of linseed, &c. for preserving the purity of light and bright pigments from the change of colour to which all drying oils are variously subject. As, however, the essential oils thus introduced weaken the body of the vehicle and occasion it to flow, so that the colours used therewith will not keep their place, and render the touch of the pencil spiritless and uncertain, they gave occasion for the introduction of *resins* and *balsams*, which give body to oils and varnishes; and the employment of resins introduced *spirituous solvents*. To these have been added *bees' and myrtle wax*,—*aqueous liquids*,—*soaps* and *salts*, as media for uniting them with oils, and a variety of dryers and other substances, too numerous to mention, with which oils, &c. have been compounded under the appellations of macgilps, gumtions, Venetian processes, and in the various empiricism of vehicles with which practice has been confounded, in endless mixture and mystery.

It is evident that amid such complicated confusion there can be no certainty of result, that the powers of chemistry are in arms, and that in such an intestine war of vehicles the best colours may be compelled to fly. In such a state of things, there can be no escape from failure and defeat—no hope but in a return to that simplicity which is a prime distinction of excellence in every art, and has marked the practice of the most eminent masters in painting.

Half a century ago the gellied vehicles which received the cant appellations of *macgilp* and *gumtion* were the favourite nostrums of the initiated in this country, and have maintained a preference with many artists to this day. These compounds of strong mastic varnish with oils rendered drying and coagulable by the salts and oxides of lead, were, according to the preceding intentions, improvements upon the simple oil vehicle, by diluting it, and giving it a gelatinous texture, which enabled it, while flowing freely from the pencil, to keep its place in painting and glazing; but their principal intention was missed by the weakening of the oil without preserving the colour or transparency of the vehicle, defects which arise from the feeble body of mastic, its softness, and a degree of disposition to darken and cloud by age. The defective colour of the *macgilp* formed with oil rendered drying by boiling or maceration on litharge, both before and after drying, was in some degree remedied by *gumtion*, composed of acetate or sugar of lead, with simple oil and strong varnish, which is subject to less change ultimately, particularly when the varnish abounds in the compound. In the using of sugar of lead, if the acid abound, which it does usually in the purer and more crystalline kinds, its power of drying is weakened, and it may have some injurious action upon colours, such as those of ultramarine and lakes. In this case a small addition of some of the pure oxides of lead, such as litharge, ground fine, will increase the drying property of the sugar of lead, and correct its injurious tendency; but too much litharge, or more than the oil will dissolve, will give a lasting opacity to the vehicle injurious to transparent colours.

Little advantage appears to have resulted from subsequent attempts to improve these vehicles, by substituting for mastic those weak resins and balsams which are but native or factitious compounds of soft resins with essential oils, similar to that of turpentine; nor from the introduction of soap, water, &c. The advantages in such cases, if any, have been merely in the working; hence many judicious artists of the present day have resorted to the use of *copal varnish,* and rejected mastic and the weak resins, contending with the difficulties of working copal for the advantages of strength, fine texture, and the greater transparency and permanence of its colour; while the resistance it opposes to re-solution by spirit, and other menstrua after drying, fits it for receiving even spirit varnishes, which may afterwards be removed without injury, and favours in other respects the texture and durability of the work: it has nevertheless the defect of cracking when it has been used without sufficient drying oil to temper it. Copal, in every mode of dissolving, swells or augments in bulk more than any other resin, like glue in water, and contracts proportionably in drying; and it is this which disposes it to crack, in which respect it is inferior to mastic.

The last operation of painting is VARNISHING, which completes the intention of the vehicle, by causing the design and colouring to bear out with their fullest freshness, force, and keeping; supplies as it were natural moisture, and a transparent atmosphere to the whole, while it forms a glazing which secures the work from injury and decay.

As it is expedient that the constitution of a picture, with regard to its materials, should be as homogeneous as possible, the proper varnish for a picture, painted with a vehicle of which mastic forms a principal ingredient, would be a mastic varnish; and this has accordingly been the prevailing practice. Anciently the mastic was dissolved in expressed oils, and not in oil of turpentine, as at present; the former, though indisposed to crack and bloom, in the manner of the latter, was nevertheless liable to become dark and discoloured by time, and difficultly removed from a picture in proportion as the oil abounded therein; in which respect copal oil varnish, though more durable than mastic varnish, is also in a less degree defective. When copal and the harder varnishes have been too soon used upon such pictures, the unequally contracting and expanding, and the various dispositions of the varnish and ground to move by heat or force, have usually cracked and damaged such pictures; but where copal has been the cement and body of the vehicle in a picture, varnishes of copal, and the harder resins, have been, for the above reason, properly preferred to mastic varnish in varnishing such pictures, and have been found less liable to chill or bloom.

But of all resinous substances in use for preparing of varnishes, the LAC of India, which is the basis of the strong and beautiful lacquered works of the east, affords the hardest, most tenacious, and durable varnish, that of *amber* not excepted. The darkness of its colour was, nevertheless, an insuperable bar to its use in painting: the discovery, however, by which this substance is entirely deprived of colour and impurities, has rendered it by far the most perfect of varnishes, and is bringing it into use in the art; so that when the difficulties usually attending the employment of new materials and means, and the obstacles common to a change of habit and practice are worn away, it will probably be the principal varnish of the painter. Yet a sudden change of established modes is neither desirable, nor to be expected from those who have waded to a safe and settled practice through a tedious experience of flattering expectations and hopes defeated.

As a general sketch of the progress of vehicles and varnishes is rather a light to show the way, than a hand to guide our practice, we will subjoin such observations of a more particular and tangible nature as experience has supplied concerning the materials of vehicles and varnishes; and, first, of the fixed or EXPRESSED OILS.

Of these, LINSEED OIL is by far the strongest, and that which dries best and firmest under proper management; but it lies under the great disadvantage of acquiring, after drying, and by exclusion from light and pure air, a semi-opaque and yellow-brown colour, which darkens by age. To obviate this as much as possible, when working with the oil alone, it is best to work the colour as stiff as may be, so as to use as small a proportion of the vehicle as may suffice; for it is a fact proved by direct and repeated experiments, that *little oil diffused through much colour is subject to little change*, and that a thin coating of linseed oil is similarly preserved by light and the oxygen of the atmosphere, as is the case in glazing or oiling out when coat-upon-coat is not applied in these ways, or a redundance of oil has not been consumed in the under painting; yet the practice of oiling out is to be deprecated when sponging with water may suffice.

Linseed oil varies in quality according to the goodness of the seed from which it is expressed; the best is yellow, transparent, comparatively sweet-scented, and has a flavour somewhat resembling that of the cucumber: great consequence has been attributed to the cold-drawing of this oil, but it is of little or no importance in painting whether moderate heat be employed or not in expressing it. Several methods have been contrived for bleaching and purifying this oil so as to render it perfectly colourless and limpid; but these give it mere beauty to the eye in a liquid state, without communicating any permanent advantage, since there is not any known process for preventing the discolourment we have spoken of as sequent to its drying, and it is, perhaps, better upon the whole that this and every vehicle should possess that colour at the time of using to which it subsequently tends, that the artist may depend upon the continuance of his tints, and use his vehicle accordingly, than that he should be betrayed, by a meretricious and evanescent beauty in his vehicle, to use it too freely. If indeed the oil were tinged with a *fugitive* transparent brown colour previously to being employed in painting, the original disposition of the oil to acquire colour by age would be compensated by the disappearance of the brown tinge given to it, so as to preserve the original freshness of the painting. Indeed linseed oil that has been long boiled upon litharge in a water-bath, to preserve it from burning, acquires such a fugitive colour; and is, when diluted with oil of turpentine, less disposed to run than pure linseed oil, and affords one of the most eligible vehicles of the oil painter.

The most valuable qualities of linseed oil, as a vehicle, consist in its great strength and flexibility; some have preferred it when *fattened by age*, or exposure to sun and air; others when *new and fresh*, or that which is *cold-drawn;* but that is the best which will temper most colour in painting, and oil expressed with a heat, which does not char or much discolour it, is equal in all respects to the cold-drawn.

For light colours and glazing the oil should be macerated two or three days at least upon about an eighth of its weight of litharge, in a warm place, occasionally shaking the mixture, after which it should be left to settle and clear, or it may be prepared without heat by levigating the litharge in the oil. This affords *pale drying oil* for light and bright colours. By a more tardy method very pale drying oil may be prepared by frequently agitating therein the smallest shot of the sportsman, or finely granulated lead, with access of air. The above mixture of oil and litharge gently and carefully boiled in an open vessel till it thicken, becomes *strong drying oil* for dark colours. As dark and transparent colours are in general comparatively ill driers, *japanners' gold size* is sometimes employed as powerful means of drying them. This material is very variously and fancifully prepared, often with needless, if not pernicious ingredients; but may be simply, and to every useful purpose in painting, prepared thus :—powder finely, of asphaltum, litharge or red lead, and burnt umber, or manganese, each one ounce; stir them into a pint of linseed oil, and simmer the mixture over a gentle fire, or on a sand bath till solution has taken place, scum ceases to rise, and the fluid thickens on cooling; carefully guarding it from taking fire. Kept at rest in a warm place, it will clear itself; or it may be strained through cloth and diluted with turpentine for use. The same methods are applicable, for the same purposes, with other oils.

POPPY OIL is much celebrated in some old books under the appellations of *oil of pinks* and *oil of carnations*, as erroneously translated from the French *oeillet*, or *olivet*, a local name for the poppy in districts where its oil is employed as a substitute for that of the *olive*. It is, however, inferior in strength, tenacity, and drying to linseed oil; and though it is of a paler colour, and slower in changing, it becomes ultimately not so yellow, but as brown and dusky as linseed oil, and therefore is in no respect to be preferred to the latter. The same may be said of NUT OILS, which have even less of the gluten or gelatine which give strength and desiccativeness to linseed oil, and come nearer to the nature of animal oils, which never dry perfectly; resembling, in this respect, the fish *oils of the seal and cod*, which, after becoming dry, soften again and even flow, and acquire deeper colour than linseed oil: nevertheless the oils of nut and poppy may be, and are, sometimes, boiled upon oxides of lead, in the manner of linseed oil, and with similar effect, in which state they may also be compounded with varnish in the formation of macgilp, and similar vehicles.

OLIVE OIL has the valuable property of permanently retaining a good colour; but this advantage is overbalanced by the almost impossibility of drying it, both which properties it communicates in mixture to other oils. Whether an oil might not be obtained of a drying quality and sufficient strength for oil-painting, and which shall have the property of continuing permanently colourless, remains for research; but, according to our present knowledge, it may be questioned whether oils do not uniformly change in colour in proportion to their natural power of drying.

The VOLATILE OILS, procured by distillation from *turpentine*, and other vegetal substances, are almost destitute of the strength of the expressed oils, having hardly more cementing power in painting than water alone, and are principally useful as solvents, and media of resinous and other substances introduced into vehicles and varnishes. They are not, however, liable to change colour like expressed oils of a drying nature; and, owing to their extreme fluidity, are useful diluents of the latter: they have also a bleaching quality, whereby they in some degree correct the tendency of drying and expressed oils to discolourment. Of essential oils, that most used in painting is the *oil of turpentine*; the rectified oil, improperly called *spirit of turpentine*, &c. is preferable only on account of its being thinner, and more free from resin. By the action of oxygen upon it water is either generated or set free, and the oil becomes thickened, but is again rendered limpid by a boiling heat, in which the oxygen is separated from it.

OIL OF LAVENDER is of two kinds,—the fine scented English oil, and the cheaper foreign oil, called *oil of spike;* these are rather more volatile and more powerful solvents than the oil of turpentine, which render them preferable in enamel painting; they have otherwise no advantage over the latter oil, unless they be fancied for their perfume. The other essential oils, such as oil of rosemary, thyme, &c., are very numerous; but it has not appeared that they possess any property that gives them superiority in painting over that of turpentine; some of them have, however, more power in dissolving resins in the making of varnishes, as is the case also with naptha or petroleum, and the rectified oil of coal tar, the intolerable scent of which latter is a prohibition to its use in any familiar work.

SPIRIT OF WINE, or *Alcohol,* is weaker and more dilute than essential oils, or even than water, and is so volatile as to be of use in vehicles only as a medium for combining oils with resins, &c.—as a powerful solvent in the formation of spirit varnishes, and in some degree as an innocent promoter of drying in oils and colours. In picture-cleaning it affords also powerful means of removing varnishes, &c.

RESINOUS VARNISHES are either *spirit varnishes, volatile oil varnishes, fixed oil varnishes,* or compounds of these, their usual solvents being either spirit of wine or alcohol, oil of turpentine, or linseed oil.

The principal VARNISHES hitherto introduced into vehicles, or used in painting, are those of MASTIC and COPAL. It is true that other soft resins are sometimes substituted for that of mastic, and that very elaborate compounds of them have been recommended and celebrated, but none that possess any evident advantage over the simple solution of mastic in rectified oil of turpentine, which is easily prepared, by digesting in a bottle during a few hours, in a warm place, one part of the dry picked resin with two parts or more of the oil of turpentine. One part of this, cleared, varnish combined with two of either of the before-mentioned drying oils of linseed, more or less according to the purpose of the artist, constitute the transparent *macgilp* of the painter. If, instead of drying oil, the simple pure linseed oil be used with about an eighth of acetate or sugar of lead dissolved in water, or ground fine, we obtain variously the opaque mixture called *gumtion,* which is of similar use, and rather less liable to change colour than macgilp. Wilson, our Coryphæus in landscape painting, used at one period of his practice a similar but simpler compound of equal quantities of linseed oil and oil of turpentine, thickened by exposure to the sun and air till it became resinous and half evaporated, to which he afterwards added a portion of melted bees'-wax; and it is probable, from similarity in the texture of their pictures, that Sir Joshua Reynolds used also the same vehicle; indeed he is said to have greatly prized linseed oil inspissated by age. According to Lanzi, Corregio's vehicle was a mixture of two parts oil with one of varnish, but of what oil and varnish we are not informed.

As other soft resins are sometimes substituted for mastic, so inferior hard resins are sometimes employed in the place of copal in the composition of varnishes celebrated as copal varnishes; but the simple solution of pure picked copal made by triturating and digesting it in oil of turpentine, as recommended by Mr. Cornelius Varley, is of more certain effect in painting and varnishing of pictures, and superior to mastic varnish in the permanence of its colours;—as a simple vehicle, it is nevertheless deficient of the tenacity and flexibility which render linseed oil so essential in painting;—combined, however, with linseed oil and oil of turpentine, copal varnish affords a vehicle superior in texture, strength, and durability to mastic and its macgilp, though in its application it is a less attractive instrument, and of more difficult management; it is, therefore, a desideratum with the artist to form a macgilp of copal, which would stand up with the flimsy firmness of that of mastic; till which can be done he must be content to mix it as above; for linseed oil is essential to prevent its cracking. The mixture of copal varnish and linseed oil is best effected by the medium of oil of turpentine; and for this purpose heat is sometimes requisite: strong copal varnish and oil of turpentine in equal portions with one-sixth of drying oil mixed together, hot, afford this vehicle; and if about an eighth of pure bees'-wax be melted into it, it will enable the vehicle to keep its place in the manner of macgilp.

AMBER VARNISH has been more reputed in painting than it merits. The process by which it is prepared is the same as that of copal; but amber is more difficult of solution, is of a deeper colour than copal, and, owing to the succinic acid it contains, dries very slowly in solution. Amber of the palest colour is most easily dissolved, and at the same time affords the best varnish.

WHITE LAC VARNISH has hitherto been only partially used in painting, for which, its being a spirit varnish that requires a warm temperature and dries rapidly, and its repugnance to combine with oils, in permanent mixture, disqualifies it by rendering it difficultly manageable as *a vehicle*. It has, however, been introduced, by rubbing, into oil colours on the palette, with a view to the giving them permanence by cloathing them, and a crispness which makes them stand up and keep their place. Its extreme transparency, and the power with which it causes colours to bear-out, have occasioned its successful introduction in the progress of a picture, to bring out and preserve the force and richness of deep colouring and shadows. To what more extensive use the ability and ingenuity of artists may apply it in oil-painting remains to be proved.

The principal recommendations of white lac, as a *varnish*, are the remarkable power and effect with which it brings out the colouring and design of a picture, and the permanence with which it so preserves them; it being neither subject to bloom, chill, nor crack, when skilfully and properly applied, according to the general rules for all varnishes, and those of spirit varnish in particular; the principal of which are a dry atmosphere and summer warmth of the apartment in which they are used. At a temperature of not less than 60°, it dries in a second or two, and coat after coat, as it dries, may be applied with a broad soft camel-hair brush. This soon becomes harder and firmer than any other varnish, and entirely free from the tackiness by which they catch and retain the dust and floatings of the atmosphere, and from the opacity and discolourment by which varnishes ultimately obscure pictures on which they are applied; so that removing this varnish from a picture becomes an unnecessary operation, though this may be easily effected, if required, by the proper use of spirit of wine. This and all other *spirit varnishes* differ from those of fixed or *expressed oils* in this respect, that spirit and essential oil varnishes are dissoluble, and removable from pictures, by alcohol and essential oils, &c.; but fixed oil varnishes are insoluble by such means, or any other not injurious to the painting itself.

Lac varnish may be combined with mastic varnish, in small proportion, so as greatly to improve it; it may also be employed in the manner of other spirit varnishes in varnishing drawings, prints, &c., which have been previously sized with isinglass; and, being first thickened by setting it to evaporate in an open vessel in a warm place, it may be passed over miniatures, &c., without previous sizing, so as to give them much of the strength, force, and durability of oil-paintings; but in all these cases it should be used in warm dry weather, or near a fire.

Upon comparing the qualities of the varnishes of mastic, copal, and lac, it will appear that the latter are successively harder and more perfect as varnishes, and in proportion to their perfection as *varnishes* is the difficulty of using them as *vehicles;* and as it is necessary that before varnishing with any of them the picture should be thoroughly dry, to prevent subsequent cracking, this is perhaps more essential for the latter than for the former.* Notwithstanding this necessity, there is one highly important advantage which seems to attend early varnishing; namely, that of preserving the colour of the vehicle used from changing, which it is observed to do when a permanent varnish is passed over colours and tints newly laid; but this it does always at the hazard and often at the expense of cracking.

This saving grace of early varnishing appears to arise from the circumstance, that, while linseed and other oils are in progress of drying, they attract oxygen, by the power of which they entirely lose their colour;

---

* In some trials, however, lately made upon oil-painting not firmly dried, with the white lac varnish, cracking of the varnish has not ensued.

but after becoming dry, they progressively acquire colour. It is at the mediate period between oils thus losing and acquiring colour, which commences previously to the oil becoming perfectly dry, that varnish preserves the colour of the vehicle, probably by preventing its farther drying and oxidation, which latter may in the end amount to that degree which constitutes combustion and produces colour:—indeed it is an established fact, that oils attract oxygen so powerfully as in many cases to have produced spontaneous combustions and destructive fires.

It is eminently conducive to good varnishing, in all cases, that it should be performed in fair weather, whatever varnish may be employed, and that a current of cold or damp air, which chills and blooms them, should be avoided. To escape the perplexities of varnishing, some have rejected it altogether, contenting themselves with oiling-out—a practice which, by avoiding an extreme, runs to its opposite, and subjects the work to ultimate irrecoverable dulness and obscurity.

The manufacturing processes of the varnishes now generally used have been recently detailed in the Transactions of the Society of Arts, &c., Vol. XLIX. p. 33., by Mr. J. Wilson Neil, to whom the Gold Isis Medal of the Society has been voted for the communication, as have also the processes for white lac varnish, in Vol. XLV. of the same publication.

CHAP. XXIV.

ON GROUNDS.

THE last thing in the order of our analysis is the ground and basis on which colours, pigments, and vehicles are applied in painting; and as the basis of fresco-painting is plaster, and that of water-colour painting is principally paper, the subject of grounds is chiefly of consideration with respect to painting in oil, in which mode a great variety of grounds have been employed, which have afforded a subject of wide speculation and experiment,—of many hopes and many failures, while the charm of Venetian art has been as fruitful in exciting the invention of grounds as of vehicles.

The subject of grounds belongs to our inquiry only so far as regards their influence upon colours, and needs no very elaborate consideration here. Among the various bases upon which grounds have been laid are the metals, stone, slate, plaster, woods, card, vellum, and cloths, in all their variety. The qualities requisite to a perfect basis are durability, infrangibility, and inflexibility, which neither of these substances comprise in perfection. Metals are durable and infrangible in the highest degree; but they expand and contract by the mere alterations of temperature, and are on this account subject to detach or throw off portions of the ground, and to craze the painting and varnish. Cloths, parchment, and paper bases, are infrangible and durable in a high degree, but very flexible, which is remedied in a measure by straining and stretching; they become therefore variously eligible bases. Wood comprises all the qualities of a good basis in a medial degree, and hence upon the whole panel affords the best basis. To treat of the peculiar qualities of all these in their various kinds, would carry us far beyond the bounds of utility, and the limits of our subject, to which the grounding of these is more intimate.

Grounding or priming is not in all cases necessary, as, for example, when stone, slate, glass, porcelain, &c., are employed, as was the case in some paintings of undoubted extreme antiquity; but when grounds are necessary, as upon metal, wood, and canvas, to be eligible, such grounds must partake all the qualities of a good basis, in being neither soft, friable, nor perishable.

The early painters in oil, being also painters in fresco, and accustomed to plaster grounds, appear to have prepared their panels, &c. with plaster or stucco, upon which they employed their colours, in some cases in water, in the manner of fresco or distemper, using size to fix them, and finishing with oil vehicles and varnish; and many such pictures have stood admirably well the ordinary effects of time, as appears among the works of Paul Veronese, Titian, Correggio, and others; but, upon cloth and flexible bases, such grounds are too stiff and friable; such bases require, therefore, a ground more of their own yielding and elastic nature, and better suited to assimilate with the materials of oil-painting, such as is afforded by tempering earths and metallic oxides with the most tenacious drying-oils, and laying them evenly upon the cloths, first coated or primed with size.

The preparing of grounds on cloths, &c. is now, however, so well performed by several of our principal colourmen, and with so much improvement, as to require little comparative attention from the artist, beyond such a general knowledge of their proper qualities and effects as may enable him to choose such as are best suited to his purpose. The colour of his ground is in like manner a matter of choice, and, generally speaking, that hue is to be preferred which partakes of the ruling colour of his picture: it was probably on this account that Titian chose to paint on a red ground, when he intended to introduce much flesh in his design, or to render red principal in his picture. It is related of the same great master, who is a prime authority in all things relating to chromatic art, that, to secure the durability and cohesion of his grounds, he imbued the canvas at the back with bees'-wax, dissolved in oil, a substance well calculated to resist damp in such a situation as Venice.

To preserve the elasticity of grounds, some drying oil should be introduced into the glue or size with which they are prepared; for the same purpose bees'-wax, sugar, treacle, albumen, &c., have been added with various degrees of eligibility and success. If the ground give way in any respect, the upper surface of the picture must fail also, and this is one of the principal causes of cracking, although by no means the only one. Those substances, however, which occasion cracking in the ground will occasion cracking in the painting; hence the importance of homogenity of process in both these subjects. Any discordance in this respect may induce cracking in a variety of ways: if a picture be painted in varnish, or even with some addition of oil, exposure to *sunshine* will inevitably crack it, by drying and contracting the upper surface, while it softens and swells the under coat upon which it is applied. *Heat* of any kind will in a less degree produce the same effect. As oils and resins imbibe moisture, *damp* will have the effect of expanding the upper surface and of cracking, blooming, and chilling soft varnishes. *Glue or animal size* in the ground unprotected will, by expanding and contracting upon damp or very dry walls, have the same effects. Thick coats of varnish, applied too rapidly, will also dispose the surface to crack by the same mechanism. Indeed a rapid drying of the upper surface before the under-painting is fixed, notwithstanding the firmness of the ground, will invariably produce cracking: this is the foundation of an artifice, of which the imitators of antiques avail themselves, by applying solutions of gum and glue over varnishes newly laid on, so as to craze the surface all over in the manner often produced by time on old pictures, &c. So powerful indeed is simple solution of gum in this respect, that, when applied upon ground glass, and dried thereon, it will disrupt and tear up the surface of the glass itself by the force of its contraction; and this is a property which belongs in a degree to some varnishes of the hard resins, such as copal, when employed over surfaces of a tenacity inferior to their own.

Other causes of cracking might be enumerated, not peculiarly attributable to the ground, such are over-stretching and mechanical violence, which do most injury to weak and inelastic substances, but against which none are entirely secure. It is apparent, therefore, that this disease of pictures, so desirable to avoid, and so often attributed to the grounds, may belong equally to the vehicles, the varnish, the pigments, or to the entire process of a painting.

It has been supposed that some grounds have impeded, and that others have promoted, drying, and that consequently the first or latter paintings have dried more or less speedily; and for this there may be some reason according with the materials of the grounds. Litharge and burnt umber are in this and other respects useful additions in the grounds. The best remedy in every case of ill-drying from the grounds will be to sponge with a weak solution of sugar of lead in water previously to painting.

With respect to the improvement of the ground of a picture, it may be worthy of experiment and inquiry, whether caoutchouc judiciously introduced upon a proper basis would not afford the best of all grounds for oil-painting?

We have only to remark, with respect to painting in water-colours, that *pure paper* is essential to the permanence of colouring,—if the bleaching acid employed in manufacturing remains ever so little in the paper, both the texture and the colours will suffer in permanence; and if, in the concern of the paper-maker to neutralize such acid, the paper be surcharged with alkali or alkaline earths, they will prove no less injurious in these and other respects: it is highly necessary therefore that these circumstances, together with the proper sizing and aluming of paper should be attended to, and that, if wanting, or if the paper happen to have been long made, the artist should reprepare it himself, by a judicious application of weak isinglass size and roach alum. And as to the practice of miniature painting, ivory and porcelain afford excellent and adequate bases entirely free from injurious action on colours.

CHAP. XXV.

ON PICTURE-CLEANING AND RESTORING.

The diseases and disorders of pictures are almost as numerous as those of animal nature, and dependent on innumerable accidental circumstances; hence picture-cleaning has become a mystery, in which all the quackery of art has been long and profitably employed, and in which every practitioner has his favourite nostrum, for doctoring, which too often denotes destroying under the pretence of restoring and preserving. The restoration of disfigured and decayed works of art is nevertheless next in importance to their production; and, as it chiefly relates to the colouring of pictures, it is a part of our inquiry with which we will close the technical portion of our work.

This medication of pictures is then no mean subject of art, but is, when divested of quackery and fraud, as honourable in its bearing as any other form of healing art; and, to be well qualified for its practice, requires a thorough education and knowledge in every thing that relates to the practice of painting, or the production of a picture, but more particularly to its chemical constitution and colouring. As, however, a picture has no natural and little of a regular constitution, it will be difficult to give general rules, and utterly impossible to prescribe universal remedies for cleaning and restoring pictures injured by time and ill-usage; we will, therefore, briefly record such methods and means as have been successfully employed in cleaning and restoring in particular cases, with such cautions as seem necessary to prevent their misapplication, confining our remarks to oil-paintings in particular.

These are subject to deterioration and disfigurement simply by dirt,—by the failure of their grounds,—by the obscuration and discolourment of vehicles and varnishes,—by the fading and changing of colours,—by the cracking of the body and surface,—by damp, mildew, and foul air,—by mechanical violence,—by injudicious cleaning and painting on,—among a variety of other natural and accidental causes of decay.

The first thing necessary to be done in cleaning and restoring is to bring the picture to its original plane and even surface, by stretching, or, if sufficiently injured to require it, by lining, which, with the transferring of pictures to new canvases, is an operation admirably well performed in London by experienced hands. In cases of simple dirt, washing with a sponge or a soft leather and water is sufficient, with subsequent rubbing of a silk handkerchief; which latter, occasionally used, is eminently preservative of a painting.

After restoring the surface to its level, and washing, the next essential in cleaning is to remove the varnish or covering by which the picture is obscured; and this in the case of simple varnishes is usually done by friction or solution, or by chemical and mechanical means united, when the varnish is combined, as commonly happens, with oils and a variety of foulness.

In removing varnish by friction, if it be a soft varnish, such as that of mastic, the simple rubbing of the finger-ends, with or without water, may be found sufficient; a portion of the resin attaches itself to the fingers, and by continued rubbing removes the varnish. If it be a hard varnish, such as that of copal, which is to be removed, friction with sea or river-sand, the particles of which have a rotundity that prevents their scratching, will accomplish the purpose.

More violent means are sometimes resorted to, but never without danger or injury.

The solvents commonly employed for this purpose are the several alkalies, alcohol and essential oils, used simply or combined. Of the alkalies, the volatile in its mildest state, or carbonate of ammonia, is the only one which can be safely used in removing dirt, oil, and varnish from a picture, which it does powerfully; it must, therefore, be much diluted with water, according to the power required, and employed with judgment and caution, stopping its action on the painting at the proper time by the use of pure water and a sponge. These cautions are doubly necessary with the fixed alkalies, potash and soda, which ought to be employed only as extraordinary means of removing spots that will not yield to safer agents. Spirits of wine or alcohol, and ether, act in a similar manner, and their power may be in like manner tempered or destroyed by dilution with water. The uniform disadvantage of all these agents is that they obscure the work, so that the operator cannot see the good he is doing, or the mischief he may have done, in the progress of his work, except by revarnishing or oiling out.

This inconvenience is, however, avoided by the safer and better mode of cleaning and removing the varnish at once by spirit of wine, tempered more or less with oil of turpentine: the practice in this case is to apply the spirituous mixture to the surface of the picture with a brush, or with carded cotton; and when, by the motion of either, the liquid has performed its office, its farther or injurious action on the design is to be stopped by another brush or cotton imbued with linseed oil, and held in the other hand; thus alternately proceeding with these tools, till the cleaning and removing the varnish is accomplished. The brushes act rather the better of the two, but the cottons imbibe the dirt and foul liquid, and are then easily exchanged for new ones. The great advantage of this method is, that the design and colouring bear out, and the progress of the cleaning is apparent.

If more action is requisite than the spirituous mixture affords, the more active essential oils may be employed, or the pure alcohol, with the addition of sulphuric ether in extreme cases; and if their action be too strong, the turpentine alone may be employed, or linseed oil added to the mixture.

Many other methods of cleaning have been recommended and employed, and in particular instances, for sufficient chemical reasons, with success; some of which we will recount, because in an art so uncertain, it is good to be rich in resources, although the legitimate doctor may deem them empirical.

In an instance of difficulty, where much care was required, we succeeded upon a picture entirely obscured by various foulness, by *varnishing* over the whole, and when thoroughly dried, removing the varnish by the above means, bringing off with it the entire foulness and original varnish of the picture, with which, in this instance, the new varnish had combined. Strong solution of gum or glue will sometimes effect the removing a foul surface mechanically, but requires care.

A thick coat of *wet fuller's earth* may be employed with safety, and, after remaining on the picture a sufficient time to soften the extraneous surface, may be removed by washing, and leave the picture pure,—and an architect of the author's acquaintance has succeeded in a similar way in restoring both paintings and gilding to their original beauty by coating them with wet clay.

An eminent artist and friend of the author passed oxgall over a very dirty old picture, which resisted washing with soap, repeating the application of the oxgall during several days, but without washing it off, till the last day, when a sponge and water easily removed the oxgall and dirt together, leaving the picture beautifully fresh and clean; the efficacy of this very safe method is due to the animal alkali contained in the gall.

Another friend, known to the public as an eminent engraver, was equally felicitous in restoring the purity of an excellent picture, by carefully washing it gradually, and in parts with some of the aqua fortis used in engraving, and cautiously sponging with water as he proceeded.

He found the acid equally efficacious in cleaning the gilding of frames.

The principle of safety in this case is, that acids, when not excessively powerful, do not act on the resinous varnishes and oils used in painting; and that nitrous acid does not act upon gold; but there is danger if the picture is cracked or abraded, both for the colouring and the canvas, and it can be employed with safety on oil-gilding only.

This method is the opposite of the alkaline process, and they may be employed together alternately in some cases to remove spots, in doing which all manner of agency must occasionally be resorted to.

Among other ingenious means of cleaning, we have it on the authority of a talented and experienced friend, that by damping the face of a picture, and exposing it to the action of a frosty night, all foulness will be effectually loosened and removed by the subsequent use of a sponge.

In every method of cleaning there is great danger of removing the glazings and otherwise injuring the colouring of a picture, which require great skill and judgment to restore.

In filling cracks and replacing portions of the ground, putty formed of whitening, varnish, and drying-oil, tinted somewhat lighter than the local colours require, should be employed; as plaster of Paris may also in some cases; and, in restoring colours accidentally removed, it should be done with a vehicle of simple varnish, because of the change of tint which takes place after drying in oil: so much is necessary, but in no case is gratuitous painting on an original picture of merit to be justified.

There is a state of declining health which occurs to every picture in the course of time, arising from the natural oil that clothes its colours and forms a semi-opaque skin, or thin surface, which, after being removed, and the picture lined if requisite, and varnished, conduces greatly to its perfect state and preservation. This operation, which gives freshness without the crudeness that belongs to pictures which have not been ameliorated by time, is necessary to every work deserving reputation.

We have thus recounted various occasions, and described a variety of methods, for the cleaning and preserving of pictures; nevertheless, we earnestly recommend that no inexperienced person should attempt to clean a valuable picture by any more powerful means than is afforded by soft water and a sponge.

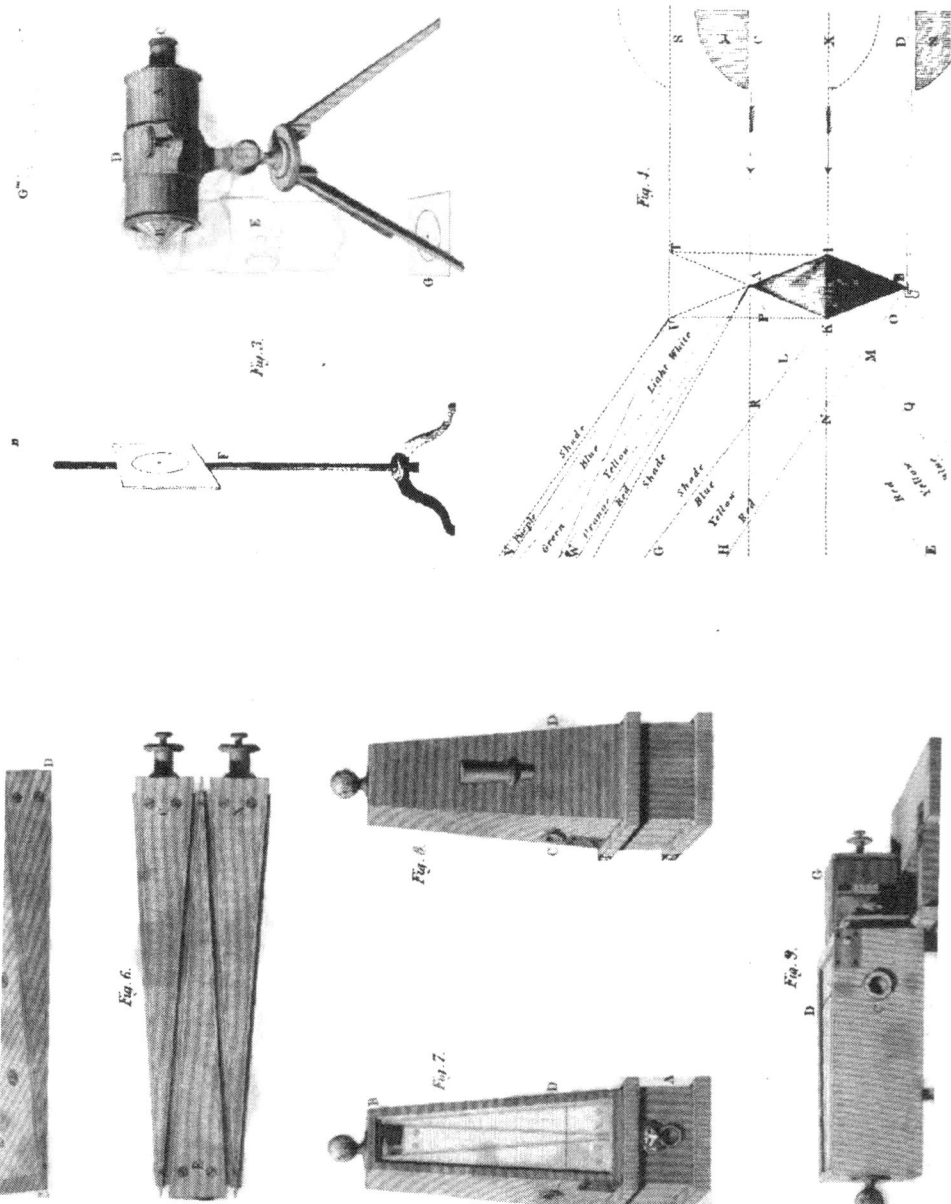

CHAP. XXVI.

DESCRIPTION OF SOME NEW OPTICAL INSTRUMENTS,

WITH EXPERIMENTS ON

LIGHT AND COLOURS.

> Although by Nature's liberal bounty bless'd,
> The fire of Genius glow within the breast,
> Collateral studies still must feed the flame
> That, clearly burning, brightens into fame.
> *Da Vinci* thus the light of science sought,
> And Art, reviving, kindled as he wrought;
> Thus *Buonaroti* rear'd his lofty name,
> And great *Urbino* brighten'd into fame;
> *Rubens* array'd in learning's lustre shone,
> And triumph'd on his allegoric throne;
> Thus, graced with all that liberal studies yield
> To form the powers of genius for the field,
> Accomplish'd *Reynolds* claims the Muse's praise.
> SHEE, ELEMENTS OF ART, p. 248.

THE principal of these instruments are the CHROMASCOPE and METROCHROME, so named from χρῶμα, *colour;* σκοπίω, *to view;* and μετρίω, *to measure.* They are optical instruments appropriate to colours, of which the author published some account several years ago;[*] and, as they have since excited some interest, are intimately connected with the subject of this work, and may

---

[*] In an essay, entitled Æsthetics, or the Analogy of the Sensible Sciences. Pamphleteer, No. XXXIII., and previously in his Chromatics, or Essay on the Analogy and Harmony of Colours.

conduce to farther instructive experience or liberal amusement, he has deemed a concise description of their construction, application, and results to be a proper appendage to this work.

His aim in these contrivances has been, by augmenting the power of the prism in the one, and adapting it to a mode of measurement in the other, to determine upon what degree of evidence the dogma of the infrangibility of Newton's homogeneal colours rested,* to illustrate the preceding doctrine of the specific powers, the relations, and harmony of colours, delivered more at large in a former work, † and to establish a standard of colours.

The common triangular glass prism has been consecrated to science by the genius of Newton as the instrument which, while it exhibits the beauties and wonders of light and colours, unfolds also the mystery of their union and separation ; ‡ it has accordingly held a principal place among the instruments of the natural philosopher, unvaried and unimproved to this day.

It is geometrically evident, notwithstanding, that as the figure of the common prism is generated by the rectilinear motion of a triangle, so it is capable of infinite variation, according to figures generated by a circular, angular, or compound motion of a triangle. Accordingly, by the motion of a triangle A B C, fig. 1, pl. II., round one of its sides A B, as an axis, is formed, as it were a circular prism, which, from its uniting the properties and figure of a lens with those of the prism, may be called a LENSIC, or LENTICULAR PRISM, or double convex PRISMATIC, or PRISMIC LENS A B C D, and fig. 4. A B I K.

Again,—by a like circular motion of a triangle E F G, pl. II. fig. 2, upon the angle F, which corresponds to the angle of refraction in a common prism, will be generated a similar double concave *lensic prism*, opposed to the above, E F G H, and fig. 4, U K I T.

The figures above mentioned are sufficient for the present purpose, yet it is evident that they are extremes, between which lie an indefinite series of

---

* Optics, Exp. v. Theor. II. Prop. II. p. 106, &c. See Exp. XII. XIII. following.

† Chromatics, or an Essay on the Analogy and Harmony of Colours.

‡ Nothing is either above or beneath the attention of the true philosopher. The mind of Newton was too great to despise even the toys of children; *soap-bubbles* blown from a tobacco-pipe, and the *prism*, long before known as a toy under the appellation of *Fool's Paradise*, became in his hand simple, yet mighty instruments of science.

intermediate figures, and that there are innumerable others, both conical and annular, generable upon the boundless variety of figure and motion. It is obvious also that *prismatic specula* may be constructed upon the same principle, which will afford by reflection optical effects analogous to those of these *lensic prisms* by refraction. We thus produce new *secondary optical powers*, the *primary* species of which are the *lens*, the *speculum*, and the *prism*, adapted to the three general habits of light by which it is *transmitted, reflected,* or *refracted;* whence arise the various powers of artificial vision, and the wonderful effects of all optical instruments. It is evident then that lensic prisms may be constructed according to all the various figures of lenses.

Of lenses there are three simple or primary kinds,—the *plane,* the *convex,* and the *concave;* of these forms are compounded three secondary species of lenses; the *plano-convex,* the *plano-concave,* and the *concavo-convex,* or meniscus; and these comprehend all the generic forms of lenses enumerated from the earliest times in books of optics. Such also are the variety of *lensic prisms.*

To facilitate the use of these *lensic prisms,* in a variety of experiments, the instrument called a CHROMASCOPE has been constructed, of which the following is a brief description.

A, fig. 3. pl. II., represents the brass tube of the chromascope, nearly two inches in diameter, and about five inches long; at one end of which is fixed, by a screw-collar, the plano-convex, or other lensic prism B; and at the other end is a small sliding tube, about two inches long and half an inch in diameter, for holding an eye-glass of seven-inch focus, or occasionally small lensic prisms; and, as a guide for the eye to the centre of the principal tube, it is fitted with a screw-cap having a small perforation at C.

The whole is held by, and slides in, the short tube, or collar D, connected with a supporting tripod having a universal joint, by which the chromascope may be turned from the horizontal position to the vertical, dotted at E, or otherwise elevated at any angle, or in any direction, for viewing objects, G, on a table, or on a portable screen or tablet F, &c.

This description of one form of the chromascope will be sufficient for a clear comprehension of the following experiments.

EXPERIMENT I.

In the centre of a white card, six inches square at least, form a black spot ⅕th of an inch diameter. Place the card upon the table G, fig. 3, pl. II., in sunshine, or a clear light near a window, and so adjust the chromascope over it, in a vertical position, that the spot may be close to, and concentrical with the lensic prism; then (having removed the lens of the eye-piece, which is unnecessary in this experiment) gradually sliding the chromascope upward, looking at the same time through the tube, the spot will appear to expand and become refracted into a beautiful annular spectrum, or aureola of the three primary colours, resembling a rainbow, as represented pl. I. fig. 1.

If now, under these circumstances, a concave lensic prism, pl. II. fig. 2, of the same refractive power of the convex prism of the chromascope be interposed between it and the object, the aureola will be, by a counter refraction, reduced to a black spot at the centre.

REMARKS.—It would be difficult to account satisfactorily for the production of colours in the above experiment by the analysis of simple light, since the coloured spectrum would vanish if the spot were removed. It is to be presumed, therefore, that the principle of shade from the spot concurs with the principle of light from the ground, by the medium of the lensic prism, in producing the circular iris. This is apparent also from the next experiment.

This concurrence of *shade* with *light* is demonstrable in all similar effects of prisms and prismic lenses in which coloured spectra are produced, although no account has been taken thereof, such phenomena having been attributed to the sole effects of light. This concurrence of shade and light is remarkable in the experiment of Newton, Optics, b. III. obs. 6, on the inflections of light and their colours, and affords easy explanation of all the experiments and observations contained in the third book of his Optics, and particularly obs. 6.

EXPERIMENT II.

If the preceding experiment be performed with a *white spot upon a black* ground, in place of the black spot upon a white ground, a similar spectrum of the same colours will be produced, in which the orders of the colours will be inverted; the *blue* in each case lying toward the *black*, and the *yellow* toward the *white;* the *red* being intermediate in each. See Chap. III. on the Relations of Colours.

REMARKS.—Various doctrines have prevailed respecting *the number of the primary colours*, there being authorities for from *one*[*] to *seven;* but the last having been the favourite number, and being sanctioned by Newton, who invented it, and supported by the apparent cogency of his attempt to demonstrate the geometrical analogy of these seven primaries with the diatonic octave of modern music, has been most generally received. If, however, the coincidence of the three colours, *blue, red,* and *yellow,* with the consonance of the primary triad C, E, G, of the musical scale, be the true foundation of such analogy; and if it be demonstrable that all other colours may be composed of these three, and that that only is primary and elementary which cannot be composed, as is the case with these three colours, then are they the only true primary colours; and as such they are recognized by the

---

[*] The late Governor Pownall maintained the doctrine of *one* only primary colour, *red.* Orange and yellow he held to be declining reds; and blue to be a privation of light, &c. Phil. Mag. vol. XXII. p. 3.

Dr. Hooke held that there were only *two* primaries, red and blue, of which all other colours are composed,—MICROGRAPHIA, p. 64.—and J. Scheffer, in his *Arte Pingendi*, 1669, distributes colours into two classes, *simple* and *mixed*, and distinguishes the first into red, blue, and yellow, thus — " Simplices colores numero sunt tres: rubeus, cæruleus, et flavus;" and adds, " Et sociabiles cunctis, Lux, id est, Albus, et Umbra, id est, Niger." § 44. p. 158.

Indeed this was the authorized doctrine of the schoolmen, and is recorded by Father Kircher, Digby, and others, previous to the time of Newton, all derived from the same Grecian source.

Scheffer treats also, under the above head, of the disagreement of the learned preceding his time, respecting the number of the primary colours—whether three, four, or five.

artist, as they were also by the antient Greeks, according to the testimony of Aristotle : * thus Homer designates—

> Jove's wondrous bow, *of three celestial dyes,*
> Placed as a sign to man amid the skies.
>
> POPE, HOMER'S ILIAD, B. XI. v. 37.

Milton too, if poets may be thus adduced, calls the rainbow—

> The triple-colour'd bow.

A late distinguished Fellow of the Royal Society has however controverted both these grounds of doctrine, and pronounced the number of the primary colours to be *four*, because, on looking through a prism at a beam of light ten or twelve feet distant in a darkened chamber, he saw distinctly that number of colours : had he chanced to have viewed the light within an inch or two of its source, and had then gradually receded, while looking through the prism, he would have discovered that his fourth primary, *green*, arose from the crossing of blue and yellow.

Had Newton too examined his spectrum near its egress from the prism, he would have perceived that his *green, orange, violet*, and *indigo primaries* arose from similar crossings of blue, red, and yellow rays; † and natural philosophers will be compelled, however tardily or reluctantly, to admit that there are in nature three primary colours only, conformably with the theory and practice of the artist.

" At the Royal Society of Edinburgh, on the twenty-first of March, a communication from Dr. Brewster was read, containing an account of a *new* analysis of white solar light. He showed that it consists of the three primary colours, red, yellow, and blue; and that the other colours, shown by the prism are also composed of these."—ATLAS, April 10, 1831.

A professor of Frankfort on the Oder has published a work to prove, in opposition to Newton, that light consists not of *seven* but of three primary colours—*red, green*, and *violet*. He remarks, that by mingling prismatic streaks of red and green, a *bright yellow* secondary is produced; by mingling green and violet a *bright blue*, &c. There is such a perverse ingenuity in this doctrine, it is founded on so singular a delusion, and is so remarkable

---

* Opp. 1629, vol. II. p. 575.
† See Note C.

an instance of the involution of truth and error, as to merit a particular exposition.

First, then, *it is true that there are but three primary colours;* but green and violet may be composed, and therefore they are not primary. It is *true again, that the green and red rays of the prism may, in confluence, produce or yield a yellow;* but for no other reason than because yellow, which is a component of green, and accompanies the warm red of the prism, is in excess or predominant in the mixture: otherwise red and green, duly proportioned, neutralize and extinguish each other, so that light would pass through or from them colourless. *It is equally true, that the green and violet of the prism mingled afford a blue;* because blue occurs in the composition of both these colours, is therefore in excess, and predominates over the neutral portion of their mingled red and green rays. Upon the same principle may the entire doctrine of this author be confuted, except only with respect to the *number* of the primary colours.

EXPERIMENT III.

It is not necessary that the objects and grounds, opposed in the preceding experiments, be black and white to produce a coloured spectrum, it is sufficient that they be *lighter* and *darker* with reference to each other; nor is it necessary that they be not coloured, since a *blue, red,* or *yellow spot,* upon a ground lighter or darker than itself, yields, in the manner above described, a coloured spectrum, as in the preceding experiments; in which, notwithstanding the particular colour of the spot itself predominates, each of the primary colours appears distinctly.

REMARKS.—The coincidence in these effects of coloured spots, with the consonances of the primary triad in every musical sound, demonstrated by Mercennus and Dr. Wallis, is remarkable. They illustrate also the natural relations, according to which the primary colours harmonize each other. See Exp. vii. and xiii. See also Chromatics, Examples xii. and xiii.

EXPERIMENT IV.

If, instead of a spot, an o, or *small circle*, be viewed with the chromascope, adjusted as in the foregoing experiments, *two* concentric annular coloured spectra, resembling the above, will appear; and if two or more concentric circles, not exceeding the diameter of the lensic prism, be so viewed, the number of the annular spectra appearing will, by an effect equally beautiful and surprising, be *double* the number of the circles viewed, in consequence of the circles being circularly refracted.

EXPERIMENT V.

That the above affords the true explanation of the double spectrum, we may be convinced, by viewing, in like manner, a narrow circle circumscribing a broad spot, fig. 10, in which case the single iris, or spectrum resulting from the spot, will appear between the two irides of the circle: there are, therefore, a double incidence and double refraction produced, the one prismatic or angular, the other orbicular or circular, whence the magnitude of the spectra of this instrument in comparison with those of the common prism.

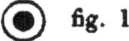
fig. 10.

EXPERIMENT VI.

If a circle, fig. 11, of not less than an inch diameter, and not exceeding the diameter of the lensic prism, be viewed as before in sunshine or a strong light, but with the chromascope gradually raised till the prism is rather more than the diameter of the circle above it, a circular spectrum will appear expanding as the instrument rises, but not *two*, as in Exp. iv., because the second iris being beyond the field of vision and angle of refraction of the instrument, never enters it. The visible spectrum of this experiment is however more beautiful and brilliant, and the primary colours more distinct and better defined therein than those of the spot and smaller circles are, owing to a more perfect refraction of the object.

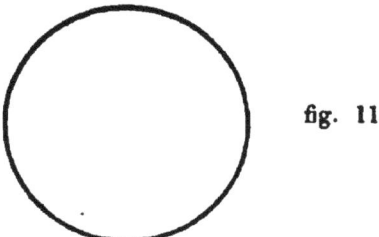

fig. 11.

REMARKS.—The spectra in these experiments will not afford perfect circles and distinct colours, unless the glass, of which the lensic prisms are formed, is perfectly free from veins, and of uniform density. Native crystal gives perfect circles, but refracts with too little power to afford well coloured spectra. Perhaps the diamond alone would yield perfect lensic prisms both for the form and colour of the spectra; and such prisms, however small, would afford beautiful aureolas, and render them effective ornaments set in jewellery. It remains also to be tried what would be the effects of these lensic prisms when constructed of other transparent substances, solid and liquid, such as salts, resins, &c.; and of Iceland crystal in particular, which has the property of double refraction, and, when formed into triangular prisms of the common shape, produces sixfold, and, by combination, other multiple refractions. This property of Iceland crystal, and other substances observed by Newton, and illustrated by Martin, remains hitherto without satisfactory explanation.

EXPERIMENT VII.

Instead of the *black* circle, fig. 11, let similar circles be formed of the purest prismatic *blue, red,* and *yellow* colours; and let each be viewed in the manner of the last experiment, when they will severally afford a spectrum each of the three primary colours; in which, nevertheless, the particular colour of the circle viewed will be predominant.

REMARKS.—In Exp. III. the same effects result from coloured spots that are herein produced from circles of single colours; but the irides, in the present experiment, are more brilliant and better defined than in the former, and somewhat better illustrations of the consonances upon which the harmonies of colours are varied and regulated.* This is not confined to the primary colours; circles of the secondaries, or any colours whatever, darker or lighter than the ground on which they are formed, afford, with the chromascope similar coloured spectra, in which the original colour of each circle becomes the archæus or key, the fundamental of a distinct harmony.†

EXPERIMENT VIII.

Let a circle of any diameter, *exceeding* that of the lensic prism, be drawn upon a vertical tablet or screen, similar to that represented at F, pl. II. fig. 3, increasing the breadth of the line by which the circle is defined in proportion to its diameter. Place this in the light of the sun, or other pure strong light, and let the chromascope be adjusted horizontally at right angles to the screen, opposite the centre of the object thereon, as in fig. 3, and at a distance therefrom equal at least to the diameter of the circle to be viewed. Then, looking through the chromascope, a beautiful coloured iris, similar to the former, will appear.

If this experiment be performed, without the tube of the chromascope, with the lensic prism alone, so that the field of vision may be extended by bringing the eye near the prism, and the object be then viewed at greater distance, an orbicular spectrum of greater magnitude and beauty will be produced.

These circular spectra will be single, because the second spectra, or irides, lie far beyond the field of the instrument; it being evident that the iris, appearing in this experiment, is similar to that which arises *within* the iris formed by the spot. In Exp. v. fig. 10, the *outward* one being lost beyond the field of vision.

REMARKS.—In the latter mode of this experiment the object may be placed upon the floor, or in any other convenient position to be viewed, as against a wall, &c., observing only that the ground of the object be of uniform colour, and of sufficient extent to render the spectrum distinct.

---

* See Exp. III. XII. and XIII.

† Chromatics, Examp. XII. XIII. XIV., &c.

In this way also the experiment may be repeated with *circles of any magnitude*, for which a screen of sufficient extent can be had, the diameter of the circle determining the distance at which it is to be viewed, and the beauty and magnitude of the coloured spectra will be proportionate. The preceding experiments, with coloured circles, &c. are of course susceptible of the same variation and enlargement.

EXPERIMENT IX.

Let C A D B, pl. II. fig. 4, represent a longitudinal section of a portion of the principal tube of the chromascope, and of its *convex lensic prism* A B I K, (described fig. 1, A B C D,) passed through the window shutter of a darkened chamber by means of a scioptic ball, Y Z.

Thus disposed, if a beam of the sun of the diameter of the tube be passed through the lensic prism in the direction X I K N, it will converge toward the point N, forming a cone of light P O N, and diverge from the point N over a cone of shade N H E.

If the light be received on a sheet of white paper at O P, where it first totally emerges from the prism, the circle of light on the paper will be bordered with *red*.

If the paper be withdrawn to Q R, the circle will be bordered with *blue*; and at the intermediate position, or focus L M, the two circles coincide without noticeable colour.

Beyond Q R the circle diverges into a ring or bow which expands in diameter in proportion to its distance from the prism upon the cone N H E. The breadth of the ring itself increases in similar proportion, and the coloured lights of which it is constituted cross each other, and diverge as represented in the upper part of the diagram, between the dotted lines U V and A W, the blue and yellow braids of light mingling and crossing at two or three feet distance from the prism.

These phenomena may be rendered beautifully visible in the atmosphere of the chamber by the steam of hot water diffused therein; as it may also by means of smoke, or powder and a powder-puff, &c.

EXPERIMENT X.

Again, let X I K, S T U, fig. 4, represent the chromascope, as in the above experiment, and T A I K U a section of the *concave lensic prism*, described fig. 2. The braid or beam of light of the diameter of the tube T I, passing through this prism in the direction C A R, will diverge from the point A, over a cone of shade A W E, forming the iris of which U V W, K F E, represents a section.

REMARKS.—These phenomena of transmitted light indicate the effects of other figures of the lensic prism; they elucidate also the powers of optical glasses in general, and throw light upon the phenomena of coloured rings observed by Sir Isaac Newton between two object glasses laid upon one another: * for the figure of *spheric lenses* may be considered as comprehending an infinity of lensico-prismatic figures, in the same manner as the circle comprehends an infinity of triangles, &c. Hence there is a double circular refraction in the incumbent lens in Newton's experiments, and a like reflection in the recumbent lens, which concur in producing the coloured rings; and the phenomena of spherical and prismatic lenses admit of similar explanation.

EXPERIMENT XI.

Admit a beam of the sun into a darkened chamber through the chromascope, in manner of Experiment ix., or by fixing its lensic prism in the scioptic ball, when a magnificent coloured iris or bow will be cast upon a screen or the walls of the apartment wherever it is directed, by turning the ball, and will be of a magnitude proportioned to the size of the room and the distance from the prism, and of a brilliancy unexampled even in the solar rainbow itself.

In the same manner, on a clear night, when the moon is in her second or third quarter, a lunar bow of faint colours may be produced.

REMARKS.—A large white screen upon a double axis, horizontal and

---

* Optics, Book II. Part I.

vertical, is of great convenience in these experiments, to receive the solar spectrum at different angles and distances. Upon the reverse of this screen, large circles and other objects may be formed or suspended.

The present Experiment affords a method by which a rainbow, of any arc, may be superinduced upon a picture into which the artist may design to introduce it, so as to try its effect, and the best way of accomplishing it.

Of that most beautiful natural phenomenon, the *rainbow*, none of the explanations hitherto offered can be pronounced universally satisfactory. The received hypothesis of the refraction of the solar rays in single spherical drops of rain, rapidly descending, dividing, subdividing, and dispersing as they fall through the atmosphere, is, notwithstanding the rich mathematical dress that envelopes it, remote from the light of demonstration. Is it not more consonant to nature and experience that the bow should be produced by one sole refraction in the entire mass of rain and the condensed atmosphere which accompanies it, than from the confusion of innumerable refractions of isolated particles or drops, neither of uniform figure, magnitude, nor position?

It is a law of optics, that when light passes from a rarer to a denser medium, it becomes refracted, and it is well known that the rainbow is produced by a partial shower, invariably opposed to the sun, and that it never exceeds a semicircle. Partial rain in its descent through the atmosphere, is progressively accelerated and resisted; hence we may infer that the shower in its fall takes a form nearly hemispherical or lensic, and that it is densest at the centre, the whole of which is favourable to the refraction and reflection of the sun's rays, in the form of a rainbow, from the entire mass of rain, in the manner of the solar bow from the lensic prism.

In like manner may be explained the beautiful colours of the clouds at the rising and setting of the sun, when the horizontal course of its rays through an atmosphere full three times the length of its vertical height, increasing in density to the centre, frequently charged with vapours, and of a hemispherical or lensic form, produces, by refracting those rays, an iris, invisible on account of the transparency of the atmosphere, but visible by the opacity of the clouds, which float or fall into the strata of its coloured lights. This is proved by the changes of colours which take place upon clouds, as they drive rapidly along under such circumstances, in windy weather. That such clouds are rarely seen of a blue or purple colour is accounted for by the blue portion of the iris lying toward shade in all cases, and in the present case falling upon the earth, or below the visible horizon, and by the blueness of the sky above.

If in the above experiment a stripe or several stripes of any opaque substance affording *shadow*, be in or upon the prism, a secondary or several concentric bows will be produced, analogous to the secondary rainbow, which is hence probably produced by a similar streak or streaks of clouds intercepting the rays which produce the original bow. These things belong, however, rather to the naturalist than the artist, to whom, nevertheless, the rainbow must ever be an interesting object, and one that is to most eyes so enchanting, that he must be worse than inanimate who is unaffected by its beauty, since it is true, as the poet has sung, in language which, though highly figurative, is yet naturally just, that when—

> Iris her lucid various bow on high
> Gaily displays, and *soothes the weeping sky*,
> The boist'rous *winds are hush'd* in deep amaze,
> And Ocean *stills his angry waves*, to gaze!

and the Iris of mythology is feigned to be the daughter of Thaumas or Thaumantias, the child of admiration!

EXPERIMENT XII.

The magnitude of the spectrum in the last Experiments renders it particularly advantageous for the performance of other experiments in or upon transient or prismatic colours. Thus;—having thrown a coloured spectrum into a darkened chamber, in manner of the last experiment, a perforated screen was placed across the most vivid part of the bow, so as to intercept it, except only at a small opening at the centre, through which a braid of coloured light was permitted to pass from the pure *blue* part of the spectrum. This *blue* ray was then received upon a second convex lensic prism; passing through which, it afforded, on a white screen placed to receive it, a second spectrum, in which, like the first spectrum, *red* and *yellow* colours accompanied the *blue*.

On varying the above arrangement, so as to produce spectra individually from the *red* and *yellow* rays of the original spectrum, they were each found to afford separately a compound spectrum, in which the primary colours, *blue*, *red*, and *yellow* were displayed, with a predominance, however, in each of the colours of the light which produced it.

The same results took place from all parts of the original spectrum.

EXPERIMENT XIII.

The opening in the perforated screen used in the preceding Experiment, being formed to give passage to smaller or larger portions of the coloured prismatic rays from any part of the spectrum, as before mentioned, by means of variously perforated slides of pasteboard, tin, or thin sheet-lead;—and the coloured lights being intercepted at their openings by semi-transparent coverings of white tissue paper, or colourless ground glass, they afforded a specific brightly coloured spot from any part of the solar spectrum cast upon them, which, being viewed through a prismic lens, exhibited a brilliant spectrum of the primary colours, *blue*, *red*, and *yellow*, whatever might have been the colour of the spot so viewed; in which, however, the ruling colour, as in the preceding Experiment, was that of the spot itself.

REMARKS.—The two latter Experiments are important in a variety of respects, both to the artist and naturalist; they demonstrate that each of the colours produced by refraction in the prismic spectrum is farther refrangible into the other colours,—all into all. That therefore the doctrine of homogeneal and heterogeneal light and colours, upon which the erroneous theory of light, and false chromatic system of the natural philosopher depend, notwithstanding the high authority upon which they are maintained,* is as contrary to fact as it is irreconcilable with the true relations of colours, and the general analogy of nature.

We therefore present this fact of the convertibility or metamorphosis of colours to the farther investigation of the naturalist, our chief purpose here being to illustrate the principles of colouring, of the relations of which in art it satisfactorily exhibits examples of invariable nature, similar to those of Experiments III. and VII. These experiments also demonstrate the indivisible triunity of light and colours, so analogous to those of the radical musical sounds, &c.

EXPERIMENT XIV.

Remove the screen employed in the latter Experiments, and let a person be placed in the solar bow at a proper distance from its entrance, so that it

---

* Newton's Opt. Exp. 5. Theor. II. Prop. II. p. 506. et seq,

cross his eyes, and the *blue* of the spectrum or bow fall upon one eye, and the *yellow* into the other, either by means of a screen or mask perforated with two openings, or even without them;—if, then, he close the eye upon which the *blue* falls, and look with the other eye toward the prism in the scioptic ball, he will perceive *a yellow light*;—then, opening the former eye, having first closed the latter, he will perceive *a blue light*;—but, finally, if he keep his position steadily, and open both eyes simultaneously, he will perceive A GREEN LIGHT only, demonstrating the concurrence of the two former colours, and both organs in the conjoint sensation of the secondary colour *green*.

This experiment extends to other cases, not only of the secondary but also of the tertiary and other compound colours, and explains some anomalies of single impression from double vision, commonly attributed to mental association, but which is hereby demonstrated to be a conjoint sensation.

REMARKS.—As far as the organs are concerned in this Experiment, the image in one eye is *yellow*, in the other *blue*, and it is by an act of the sensorium they become GREEN:—so in the erect view of the inverted image which appears on the retina in vision, that has so much perplexed philosophers, it is by an act of light reflected from the object that the image is painted inverted on the retina:—it is by another and perhaps similar act between this image and the sensorium that the cognition of the true and erect position of the object arises. It is the same as to number, motion, and magnitude—*two* images are painted, one on each retina, but a *single* image only is cognised, if the one eye receive a larger image than the other, as is sometimes the case in imperfect vision. Viewed alternately with one eye, the object will appear large, with the other small; while viewed with both eyes together, the object will appear of a mean magnitude. The same may be observed upon using a pair of spectacles, of which the two glasses have different foci, and consequently different magnifying powers.

We have seen by the present Experiment, that if light (or any white or light object) be viewed through a *blue* medium with one eye, and at the same time through a *yellow* medium with the other, the conjunct sensation of such object will be *green*; and in this way all colours may be compounded in the sensorium itself:—and if after some time thus viewing a white object as *green*, the coloured media be removed, the object will appear *red*, according to the law by which ocular spectra are produced; nevertheless, if the object be viewed by that eye alone to which the *blue* medium had been applied, the spectrum would not, according to the same law, be *orange*:—nor if viewed by the eye alone to which the *yellow* medium had been applied, would the spectrum or object be purple; but would in both cases be RED, demonstrating that ocular spectra belong to the *sensorium*, and not to the *organ*—that in vision the sensorium is active, and that the return sensation through each eye will be *compound*, although the exciting cause, with respect to either eye, has been simple: i. e. both the eye affected by *blue*, and that affected by *yellow*, would singly and alternately see *red*, instead of orange in the one case and purple in the other: and, in like manner, according to this new analogy, in all other cases of vision through coloured media.

EXPERIMENT XV.

Let an assistant be placed, as in the last experiment, in the broad spectrum of the lensic prism, so that the *red* fall distinctly in one eye and the *green* in the other; if, then, each eye be alternately shut while the other remains open, the colour respectively shining upon each organ will be seen alternately as before; yet, if both eyes be opened together, NO COLOUR WHATEVER WILL APPEAR.

By means of a small mirror held in the hand, the experimenter may perform these Experiments upon his own eyes, without an assistant; but in either case the spectator should be placed at a distance from the lensic prism, to be governed by the breadth and colouring of the spectrum.

REMARKS.—This Experiment extends, like the preceding, to other cases, and demonstrates in a new way the neutralizing, extinguishing, complementary, contrasting, or compensating powers of colours, which may justly be considered as the key to chromatic science.

Mr. Smith, of Fochabers, has lately published a very pleasing Experiment, which discloses a new mode of ocular excitement, producing coloured spectra:—Mr. Smith (says Sir David Brewster [*]) states, that when a candle is held near the right eye, so as to be seen by it, but not by the left eye;— and then when both eyes look at a narrow strip of white paper, so as to see it double, the image of the paper seen by the right or excited eye, will be *green*, and that seen by the left, or eye protected from the candle-light will appear *redish*. On trying this experiment with a strip of paper, nine

[*] Lond. and Edin. Phil. Mag. Vol. I. p. 249. 343. Vol. II. p. 168.

inches long and the tenth of an inch wide, we observed the effect described to arise gradually;—when immediately removing the light from the right to the left eye, we found the colours at first to remain as above unchanged, but gradually they declined, and both images became as *colourless* as the light;—they then again gradually acquired colour, that which had been *green* becoming *red*, and that which had been *red* becoming *green*. It is evident, therefore, that these colours depend upon opposed spectra *produced in time*,—by the *hot* light a green spectrum, and by the *cool* compensating shadow a *red* one, and that they are equivalent and complementary;— evincing also our doctrine of the excitement and exhaustion of the principles of light and colour in the organ of sight; whence it may become a useful practice in painting, that the artist view his colouring with his right and left eyes alternately closed, so as not to be subject to error by the false excitement of either organ.

That the colours *green* and *red* are equivalent and complementary in the above Smithian Experiment, may be farther proved by gradually turning the strip of paper from the perpendicular to a horizontal position, when the *double* object will merge exactly into the *single*, and the two colours coincide and become white as the light.

In whatever way this neutralization of equivalent colours is produced, there is a union or coalescence of the primary triad, blue, red, and yellow, in due subordination; and it is remarkable, that when it arises from the union of two colours, as in the present Experiment, these colours bear the relation of that interval in music which is called the *fourth*, corresponding to the *diatessaron* of the Greeks, which was held by them to be the concord upon which all others depend. This harmonizing power of colours renders it evident that it was not without some foundation Father Castel is said to have attempted to construct an ocular harpsichord, for the purpose of exhibiting to the eye a pleasurable sensation from colours, analogous to that which the ear receives from musical sounds; but he evidently could not have succeeded, because his scale was erroneous, and because the harmonious effects of colours are not, as he treated them, merely successive or *temporary*, like those of sounds, but they are simultaneous, co-expansive, and governed by *space*.* In these Experiments, properly coloured glasses, or other media, such as transparent coloured liquids, may be substituted for the colours of the spectrum, and with the same results.

* Exp. XXVI.

EXPERIMENT XVI.

Let a solar bow be cast upon a white screen in a room nearly darkened in manner of Experiment xi. at a distance of ten or twelve feet from the lensic prism, at which distance the secondary colours, orange, green, and purple, are apparent. Then hold a small flat ruler across the iris, so that it cast a shadow upon the screen, when it will be found that the space of the shadow which displaces a portion of the bow will be supplied with an inverse spectrum, the colours of which will be perfect contrasts, colour for colour, to those of the bow ; thus, in place of *purple* in the bow will be *dark yellow* in the shadow, in place of *green* will be *dark red*, and in place of *orange*, dark *blue;* and so it will be with other contrasts and colours of the bow, taken nearer its source with a proportionably narrower ruler.

REMARKS.—This experiment is important to the artist by demonstrating that *shade* is in all cases a contrast to *light*, not only in effect or power, as chiaroscuro, but also in colour.

It corrects also the error into which the naturalist has fallen in explaining the phenomenon of the blue shadows which occur with the orange light of the rising and setting sun, or other warm-coloured light, noticed by Leonardo da Vinci, Counts Buffon and Rumford, and attributed to *blue* reflected from the sky, which, in truth, is merely complementary to the *orange*, or golder colour of the light; for if, at any time of day, any colour be given to the sun's light, by passing it through coloured glass into a darkened chamber, the shadows of such light will always be of the colour which contrasts it, notwithstanding the reflex blue of the sky ; accordingly, if in a bright sunny day, when the sky is bluest, the shadows of an object be projected on white paper at a window opposite the north, into which the sun never enters, such shadows will be so far from blue, that they will be of colour more or less warm in proportion to the blueness of the sky. The same effects are also uniformly produced by the colours of artificial light.

Another remarkable circumstance of the *coloured shadows* of our Experiment is, that those of a colour related to light, such as yellow or orange, will appear *lighter* than their lights, which are purple and blue, and vice versa of those coloured shadows whose colours are related to shade, evincing physically the truth of such relations, by which some colours appear to carry light, and to *advance*, while others carry shade and *retire*, according to the principles of painting ; and interpreting also the beautiful compensation of nature by which shade increases with light, and vice versa.

EXPERIMENT XVII.

Upon viewing, in the manner of Experiment 1. a black spot, three-quarters of an inch in diameter, upon a white ground at about three inches distant from the lensic prism, a beautiful *blue* circle, inscribed with white, and circumscribed by black, will be produced; and if a similar white spot, upon a black ground, be viewed in the same manner, a *red and yellow* circle, inscribed with black and circumscribed with white, will appear. By varying the colours of the ground and spot in this experiment, circles of any required colour may be obtained; thus a *black spot* upon a *yellow ground* will yield a circle of *green*, &c.

EXPERIMENT XVIII.

If any number of concentric circles, variously distinguished by figure or colour, circumscribing a spot in manner of Experiment v. fig. 10, be viewed in the same way, a compound spectrum of the like number of circles will be produced within that of the spot, succeeding one within another as the instrument rises, but in the inverted order of the object, the reason of which has been given.

EXPERIMENT XIX.

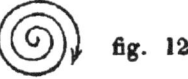

fig. 12.

Let a spiral, fig. 12, be formed of any number of involutions; then adjust the chromascope to its centre, as in Experiment 1. and, looking through the instrument, gradually elevate it, when the spiral will appear refracted into *two involved spiral irides;* and if the figure be barbed, or arrow-headed, at either end of the spiral, a like arrow-head will be found at the central end of one of the involved irides, and another at the external extreme of the other, the reason of which is apparent from the double incidence and refraction adduced, Experiment V.

EXPERIMENT XX.

Again; if a spiral of any diameter exceeding that of the chromascope be viewed at a proportionate distance, as in the last Experiment, a *single* spiral iris only will be seen, as in Experiment VI., and if a head be formed at the outward extremity of the figure, it will be seen at the centre of the spectrum; and if the figure be drawn to represent a serpent as large as the boa constrictor, it will seem as it were to uncoil itself as the instrument recedes from it, and will be beautifully variegated with prismatic colours.

EXPERIMENT XXI.

On a clear evening, when the moon is at the full, remove the small tube of the chromascope, and so adjust it with the open end toward the moon that it may be viewed through it by placing the eye close to the lensic prism, when the moon will be refracted into a beautiful lucid orb, the colours of which will not be at all inferior in brilliancy to those of the solar spectrum, Exp. II.

A plano-convex, or conical lensic prism, with the plane toward the eye, is the most convenient for close vision, and the eye being close to the plane, and the apex toward the object concentrically, a wider field of vision is commanded.

EXPERIMENT XXII.

On a dark night, when the sky is clear, the planets and fixed stars may be viewed in manner of the above experiments, when the light of either will be refracted into coloured orbs, differing from that of the moon only in the breadth and brilliancy of its colours; proving that the light of the heavenly bodies differs not by the analysis of refraction.

In the same manner may be examined the lights of the glow-worm, lightning, phosphori, &c.; but when these lights are faint, the yellow is difficultly distinguished in their spectra.

EXPERIMENT XXIII.

If day-light be admitted into a darkened chamber through a small round aperture in the window-shutter, or a sun-beam be received on a ground glass, or tissue paper, which covers the opening; and it be then viewed, as in the last experiments, at any convenient distance, a brilliant circular iris will be produced, as in those experiments. For this purpose a circular card, or piece of thin sheet lead, or other metal, perforated with the aperture required, and fitted to the socket of the scioptic ball, is well adapted.

REMARKS. — In the same manner many of the preceding experiments, by reflection from opaque and coloured objects, may be performed by means of figures perforated in the manner of stainfoils, in card or thin sheet metal, and adapted to the socket of the scioptic ball, or opening in the window shutter; and these experiments have the advantages of superior brilliancy, &c.

EXPERIMENT XXIV.

Into the short small tube, or eye-piece of the chromascope, fit a small lensic prism, of a plano-convex or other required shape, at the interior extremity. At the opposite, or prism end of the principal tube of the chromascope, fit a plane white ground glass, perforated metal plates, coloured glasses or lenses, &c. as may be required: or the whole may be managed by a sliding apparatus, fixed at the end of the chromascope, in manner of the magic lanthorn.

Thus provided, the instrument will be adapted for a variety of the preceding and other experiments in a light apartment, in the sun, or in artificial light,—the large tube supplying the place of a darkened chamber or screen. Thus, e. g. to demonstrate the refrangibility of the coloured rays of the prismic spectrum, as in Exp. XIII., having fitted the large tube of the chromascope with a colourless ground glass, and a perforated plate, cast either of the original pure colours of the prismic spectrum on the object glass, or perforated plate at the end of the chromascope, and view it through the lensic prism of the small tube at the opposite end, &c.

GENERAL REMARKS.—To describe all the experiments to which these instruments conduct, whether for use or amusement, is impossible. The foregoing are sufficient to show some of their applications and to indicate others. What is therein performed by the refraction of lensic prisms may, in many instances, be accomplished by the reflection of similar specula; add to which the other figures of these prisms, before pointed out, a similar variety of annular prisms and their combinations, together with the variations of which the chromascope itself is susceptible, from the simple hand-glass to its combinations with other optical instruments, and a new and extensive field is opened to the ingenious enquirer for dioptrical experiment, adapted equally to instructive amusement and the advancement of science. Of other applications of these instruments none is more obvious than the facility with which they may be adapted to the magic lanthorn, so as to exhibit their effects by artificial light.

It remains yet that we describe the METROCHROME, the other instrument alluded to at the commencement of this chapter, and illustrate some of its principal uses and effects by experiments.

ON THE METROCHROME;

OR,

STANDARD OF COLOURS.

An accurate and determinate mode of measuring and denominating colours, so as to convey precise ideas of their hues, shades, and relations, has hitherto been a *desideratum*, not only in fine art, but also to the chemist and geologist, the botanist and anatomist, the optician and astronomer, and in every department of natural philosophy. Nor has it been less desirable in commerce; to the cultivator, and to the manufacturer, not to enumerate also the many utilities and appliances of such a standard in various other concerns of taste and science.

In painting, in particular, it has the important office of establishing definitively the proportional powers of colours upon which their equivalence, or faculty of harmonizing each other in every possible case depends, the accomplishment of which has been among the principal objects of the METROCHROME described, and so applied, in the following experiments.

EXPERIMENT XXV.

Fig. 5. Pl. II. represents part of the above-named instrument, in which A B C D is a hollow prism or wedge, in each side of which is cemented, and secured by a brass frame and screws, a colourless plate glass $efgh$, which glasses touch each other within at the end eg, and diverge or separate at the other ends to the thickness of the wedge at fh. We thus obtained a hollow prism, pervious to light and vision, which might be filled with a transparent coloured liquid by means of an opening and screw stopper in the end B D. To prevent compression of the liquid the stopper should be perforated lengthwise, that the air may escape, and the perforation be secured by the screw or plug E.

It is evident this wedge, being so charged with a coloured liquid, and viewed opposite the light, will, throughout its broad face, present a perfect gradation of colour, from the utmost diluteness or minimum, at the convergent extreme $g\,e$, where the glasses touch, to the utmost depth or maximum at the divergent extreme $f\,h$, where they are at their utmost separation.

On one side of the wedge, C D, is screwed a brass scale of the exact length of the cavity within the glasses $e\,f\,g\,h$, geometrically divided into 32 degrees, each subdivided into four others, forming evidently an accurate measure of thickness, increasing at each division, from the point of contact of the glasses, and are consequently also a true numerical measure of the intensities of transparent colour throughout the wedge. Such a prism, charged with a *blue* liquid, will form a *cyanometer*, or measure of blue ; with a *red* liquid, an *eruthrometer*, or measure of red ; and with a *yellow* liquid, a *xanthometer*, or measure of yellow, the colours of such liquids being adjusted to a given intensity.

EXPERIMENT XXVI.

Again C E X, fig. 6. Pl. II., represent in combination three of the above described prismic wedges, or colour gauges, accurately formed and graduated to the same scale, and having prepared three liquids of the three true primary colours of equal powers or intensities,* charge the cyanometer C with *blue* liquid, the eruthrometer E with *red*, and the xanthometer X with *yellow*.†

The gauges thus prepared may be combined in pairs alternately to produce the *secondary* colours *purple, green,* and *orange,* similarly gradated of all *shades ;* or, by changing them endwise, of one uniform shade of all *hues,* and

---

* This is as easy to a correct and practised eye as the tuning of musical strings is to a musician ; nevertheless mechanical aids may be resorted to. E. g. Intercept the light passing singly through the coloured liquids at any given division of the wedges by three equal bars, and receive their shadows on white paper at equal distances, varying the coloured liquids till they cast equal shadows, or till they are lost at equal distances ; or various other photometrical contrivances may be employed.

† The liquids may be, for a *blue*, a solution of sulphate of copper; for the *red*, liquid rubiate diluted with water; and for *yellow*, aqueous tincture of saffron, or terra merita.

the proportions of the compound of any hue will be denoted by comparing the numbers opposed to it on the two scales.

By similar management of the three gauges may be produced of the same originals, blue, red, and yellow; the *tertiary* colours, *olive*, *russet*, and *citrine*, and all other compounds, with a like power of measuring and computing their proportions.

Thus used in conjunction the three gauges constituted a METROCHROME, or general measure and standard of colours.

REMARKS.—It has been suggested by a friend intimately conversant with the philosophy of colours, " that the most effectual mode of making a standard would be to have the wedges filled with coloured liquids so diluted, or at angles so acute, that a ray of light transmitted through either of the wedges at the centre, taken as zero, should, as nearly as possible, correspond with the red, blue, or yellow ray from the prism at any given distance. This would become a standard if the centre gave the tint in truth and just intensity at *zero*, and the wedges always had the same length and angle. These records of corresponding quantities above and below zero would become standards of tints amounting, in a scale of only thirty-two degrees, to nearly three hundred ternary tints, independent of the binary, and the degrees of intensity of the primary."

EXPERIMENT XXVII.

For a convenient mode of managing the graduated wedges in conjunction, a tube or case, fig. 7. Pl. II., has been adopted, consisting of a pedestal A, which forms a foot or stand for the instrument when in use, and an obelisk B, to receive the wedges. The latter has a glass front through which the graduated sides of the wedges may be seen, and the sides of the obelisk are perforated with two small openings opposite each other, C D; through which, and the wedges altogether, a direct view may be had, or a beam of light be thrown. The obelisk is connected with its pedestal by hinges, enabling it to fall back at right angles therewith; in which situation it is further supported by a falling leg E, in the back view of the apparatus, fig. 8, as appears in its horizontal position, fig. 9.

In using the metrochrome, place it in its horizontal position across the light to be viewed or transmitted; turn out the lid of its pedestal F, for the wedges, charged with their respective colours, to slide on, into, or out of the obelisk, as appears at G.

Parallel with the axis of vision, and over the centre of the two openings in the sides of the metrochrome C D, is an index formed by a fine wire stretched across,—or better by a line drawn with a diamond pencil across the face of the front glass. If now, for example, the blue gauge be slid in, till its scale reaches 24 degrees beneath the index, the red gauge till it reaches 15°, and the yellow till it reaches 9°—or any similar proportions— a beam of light being cast through the whole by the openings C D, will pass achromatic or colourless, provided the coloured liquids with which the gauges have been charged are of pure colours and equal intensities. The light viewed through them in the opposite direction will also, of course, appear colourless. It is not necessary that the blue gauge be placed at 24° in this Experiment;—any other position of the scale will equally afford the achromatic compound of yellow, red, and blue, upon properly adjusting the other two wedges to such other position, when it will invariably result that the proportions of the three are as 9, 15, 24, or in the ratio of 3, 5, 8, in remarkable coincidence with the geometrical analogy, or proportional intervals, of the *common chord*, or harmonic triad of the musician, which is the foundation of all harmony in sounds, as deduced by Tartini and others from the string trumpet and monochord. In the three primary colours combined thus in achromatic unity, or accordance, the power of yellow is as 3, that of red as 5, and that of blue as 8.

EXPERIMENT XXVIII.

Having adjusted the colour gauges in the position of the metrochrome fig. 9. as in the last Experiment in the complementary proportions $3°+5°+8°=16°$; if the blue be then withdrawn, a *perfect orange*, composed of 3 yellow and 5 red will remain, and be visible either by transmitting a beam of light, or looking through the gauges in the metrochrome, as before described; and this orange of 8° is consequently the equal contrast or equivalent complementary of blue of 8°; 16° being the amount of neutrality.[*]

---

[*] This numerical equivalence of orange and blue is as remarkable as the corresponding intervals in the diatonic scale of the natural major mode in music, which are both half-notes.

If now the blue gauge be restored to the exact position, from which it was removed, in the metrochrome, and the red gauge be withdrawn in its stead, a *perfect green*, composed of yellow of 3° and blue of 8°, will remain visible in manner and place of the orange; and this green of 11° will be the equivalent of red of 5°, their numbers amounting also to 16°.

Finally, by restoring the red gauge to its position in the metrochrome, and withdrawing the yellow gauge, a *perfect purple*, composed of red of 5° and blue of 8°, will remain in like manner; and this purple of 13° will be the equivalent of yellow of 3°, which again also amount to 16°.

REMARKS.—From these data may be deduced the relative powers of other colours, for the tertiaries are regular compounds of the secondaries, as the latter are of the primary colours, as denoted by the following Table:—

		Primary.							
		Blue	Red	Yel.					
	16° Neutral ..	8°	5°	3°	White and Light.				
Secondary.	Orange . 8°=	0	5	3					
	Green .. 11°=	8	0	3					
	Purple .. 13°=	8	5	0					
	Neutral	16	10	6	Grey.				
						Blue	Red	Yel.	
Tertiary.	{ Orange .. 8=	0	5	3	}	=8	5	6	Citrine.
	{ Green .. 11=	8	0	3					
	{ Orange .. 8=	0	5	3	}	=8	10	3	Russet.
	{ Purple .. 13=	8	5	0					
	{ Green .. 11=	8	0	3	}	=16	5	3	Olive.
	{ Purple .. 13=	8	5	0					
	Neutral	32	20	12	Black.				

It appears above, that the amounts of the primary colours which constitute the secondaries are proportionally 16°=10=6; and that those of the secondary colours, which constitute the tertiaries, are 32=20=12; but both

CHROMATIC EQUIVALENCE.   249

these are the same in ratio, as 8=5=3, or that mixture of the primaries which compensate or neutralize each other equivalently; accordingly the neutrals, black, white, and grey, may be compounded of secondary, primary, or tertiary colours, as we have shown elsewhere.

	Blue.	Red.	Yel.	
If we add to	16	5	3	which are the constituents of *olive*,
the		5	3	which constitute *orange*,
they amount to	16	10	6	which constitute the neutral *grey*: accord-

ingly orange is the contrast and complementary of olive.

	Blue.	Red.	Yel.	
So if we add to	8	10	3	which belong to *russet*,
the	8	0	3	of green,
we again obtain the	16	10	6	of grey;—and green is the contrast of

russet.

	Blue.	Red.	Yel.	
And if we add to the	8	5	6	of citrine
the	8	5	0	of purple, we
also obtain	16	10	6	of grey; and purple is the contrast of

citrine. The tertiaries have, therefore, similar relation to black or shade that the primaries have to white or light. Hence the relations and proportionals of colours terminate circularly, or at the point of commencement. They are therefore complete, and admit of systematic arrangement.

Accordingly we have constructed the Scale fig. 2. Pl. i. upon the principle of which may be deduced the equivalent relations of colours to infinity. This scale comprehends six circles, constituting three stars, each having six radials, or branches, with black at the centre; but of which we need not repeat the description already given in Chap. iii., and of which we have now supplied the principle and rationale.* Any one so disposed may refine mechanically upon this plan, by giving motion to the outward scale of figures, so as to work with the inward fixed scale; all possible relations and combinations of colours numerically, and in other ways. It is evident that all colours and contrasts may be determined upon the same principle of mensuration, so that the artist may not only know and name his hues, but also form a correct judgment of the proportions in which they will neutralize, and the quantities either of surface or intensity in which they harmonize

* Note D.

each other;—and carrying therewith a fine eye to his performances, he may the more readily satisfy the demands of a pure and cultivated taste. All the rules of poetry will not, however, make a poet; nor will the utmost refinements of mechanism or science give genius to the painter, however they may correct and aid him as important means in his art, and help him to realize the conceptions of a poetical imagination.

To treat of the innumerable uses to which the metrochrome may be applied in the various affairs of life and literature, would be little to the purpose of the present work—and they are perhaps sufficiently obvious. To give to the instrument itself the utmost simplicity and perfection, would render a much more important service to art. And as the triangular prisms, first employed by natural philosophers, were formed of glass planes and filled with water, &c., ere prisms of solid glass were constructed, so it is perfectly practicable to form coloured gauges of solid glass to supply the place of the hollow wedges of the metrochrome. For this purpose it may be suggested, that clear flint glass may be tinged by the sulphuret of gold of a *red* colour— by the sulphuret of silver, *yellow*—and by the sulphuret of cobalt, *blue*; or, otherwise, by the various precipitates of these metals, or the same metals minutely divided.

It is not necessary that massy coloured glass wedges should be formed for the above purpose; on the contrary, a new mode of perfect adjustment of the colours of the glasses would arise if wedge-shaped pieces of flat glass of the eighth of an inch uniform thickness, polished at the edges, were employed—because the glasses, being of uniform thickness, might be chosen of equal intensities, and a perforation of the obelisk of the metrochrome, for light and vision, of the diameter of the thickness of the glass would be sufficient for all the purposes of the instrument; and these means would render its construction as easy as that of most other metral instruments.

The following ingenious mode of obtaining the results of these instruments has been suggested :—" Instead of transmitting the light, as required, through two or three such wedges at once, let them be so adapted to a whirling disk, that each wedge may be brought to pass in rapid succession before an eye-hole, through which the light transmitted through the wedges will present compounded tints of different hues and intensities, by sliding the wedges so that different thicknesses thereof may pass before the eye-hole. In this way every possible tint may be compounded, and the known quantities of its constituent colours determined by the scale."

The same may be accomplished in other ways, as by uniformly thin transparent plates of the required primary colours with which hues and shades may be compounded, and the proportions of the compounds denoted by the number of plates employed; and also mechanically and chemically, as in the following Experiments:

EXPERIMENT XXIX.

Divide a circular white card of three inches diameter by a line across, into two equal semicircles, one of which paint of a light blue colour; again, divide the other half-circle from the centre, in the proportion of five to three, and paint its larger portion of a light prismatic red colour, and the smaller portion of a bright yellow, so that the three colours shall be of equal pureness and intensities. The face of the card will then be divided and coloured in the proportions of three yellow, five red, and eight blue. Pass a pin through the centre of this card, which spin rapidly thereon with its face toward the light, when the three colours of the circle will blend and vanish, and the card will be seen of a dull white colour, which is most apparent when contrasted by some dark object behind it.

EXPERIMENT XXX.

Three cylindrical glasses, of equal diameters and equal height, were filled with aqueous tinctures, the one blue, another red, and the third yellow, of pure colours, and reduced by diluting with water to as nearly equal intensities as the eye could judge. Then into a glass tube, nine inches long and half an inch in diameter, accurately graduated lengthwise, portions of each of the above coloured liquids were poured alternately till the mixture upon shaking became neutral or colourless. Upon measuring the quantities of the coloured liquids remaining in the cylindrical glasses, the proportions wanting were eight blue, five red, and three yellow, which formed the proportions of the neutral liquid in the tube within a fraction.

This Experiment was repeated several times with the same result, by pouring the coloured liquids into the graduated tube in the above proportions, and shaking;—more of either colour then added to the mixture gave its own hue. The colours of the liquids should not be so deep as to disturb their transparency; the eye is also a better judge of the hues and intensities of light or pale than of deep or dark colours.

EXPERIMENT. XXXI.

The last experiment lies under the disadvantage of the permanent combination of the three liquids, so that we cannot be assured that the disappearance of the colours in the mixture is not the effect of chemical change: if, therefore, we were to tinge clear saturated aqueous solution of *potash* of a *blue* colour,—*oil of turpentine*, or other colourless oil, of a pure *yellow* colour, and *alcohol red*, and, with these coloured liquids, repeat the above Experiment, the neutral compound being left at rest, would soon separate, and the three liquids detach themselves in their proper colours, demonstrating at once the proportional powers of the primary colours, synthetically and analytically.

This Experiment may be adapted to the various other compounds and relations of colours; e. g. for exhibiting the contrasts complementarily, by variously tingeing these liquids as the case may require.

REMARKS.—A tube employed as in the above experiments becomes a MONOCHROME, so to call it, in which the three primary colours are united harmonically, under certain invariable natural proportions, analogous to the harmonics of the *monochord and trumpet marine* of the musician, upon which simple instrument the natural theory or science of his beautiful art is grounded;[*] so also may be developed, by this monochrome, the entire scale of colours analogically as numbers to unity.

Notwithstanding the simplicity of this latter instrument, the metrochrome has a wider range, and many obvious advantages of which the monochrome is deficient; and this in particular, that a more perfect neutralization of the colours can be effected without even a suspicion of chemical change by the action of contact and mixture in the liquids: these instruments, however, help to confirm each other's results by coincidence.

To the same end other means may be resorted to, and such also as are not liable to chemical exception, by mingling *measured* quantities of *coloured*

---

[*] See Stillingfleet's Principles and Powers of Harmony, p. 22, § 35.

powders of equal purity and intensities, or by *numbers* of *coloured glasses*, or other more tenuous transparent coloured substances, of equal thickness, pureness, and intensities; by which the general laws of chromatic accordance may be established upon a wider induction.

It may perhaps be objected against the metral accuracy of these instruments that they have not an indisputable or invariable standard; but the same objection holds against the barometer, thermometer, hydrometer, and all other meters, the most perfect of which afford but approximations to truth; and this, in the present instance, is all that can be of use to the artist; and relative perfection is all that we claim for the metrochrome.

The eye must be the judge and test of purity and depth of colour, as the ear is of sounds; and as many musicians are to be found who could tune strings and instruments in respect to pitch, tune, and temperament, to the acquiescence, if not entire satisfaction, of every good ear; so few artists, perhaps none, would dispute the relative depth or purity of any two or more colours equalised by a good eye; and upon this principle depends the adjustment of the gauges of the metrochrome, though other aids may be resorted to if required. This instrument is therefore qualified in principle and practice to become a universal standard of colours by which the philosopher, the artist, and the merchant may register and communicate accurate records of colours from the most distant places, and from age to age.

CONCLUSION.—The foregoing experiments and observations, selected from our occasional and desultory investigation of this delightful branch of optics, might have been greatly extended had it been the leading subject of this work; but as it may be thought by some that we have already passed the bounds of a proper subordination, we will add only a few words more, and have done.

The chromascope, it is evident, is capable of great variation in its construction and application; and much remains to be done in the formation and adaptation of the lensic prisms. When solid wedges of coloured glass of pure hues and adjusted intensities shall have been constructed, they will contribute to the simplicity, permanence, and perfection of the metrochrome. For these instruments also a properly constructed chamber, resembling the magic lanthorn, for employing them with artificial light, should be constructed, all of which would improve their application and facilitate research in the hand of an ingenious inquirer.

CONCLUSION.

Upon the whole, it has appeared that whether we experiment upon the *inherent or reflected colours of pigments, &c.*, the *transient or refracted colours of prisms, &c.*, or the *transmitted colours of transparent liquids, &c.*, they present the same uniform relations, and in this illustration our design in these experiments terminates; nevertheless we have not ceased to intend, during the last twenty-five years, to pursue this inquiry in such a way as might carry these instruments to such relative perfection as might qualify them for general use, according to their objects; but, having satisfied our primary intention by these experiments, our ultimate design has given way from time to time to more urgent occupations; but the field of research herein is still open, and, to any person having leisure and inclination for such inquiry, there is ample room for the exercise of skill and ingenuity, with a fair prospect of beautiful and useful results; indeed, an able artist, and most ingenious gentleman, the late Mr. Stephens, formerly one of the professors of the Royal Military College at Sandhurst, had, some years since, taken up the subject with zeal and assiduity from our hands; and we had reason to expect very interesting communications from one so well prepared for the research, when his sudden death, in the cathedral at Exeter, unfortunately put a period to this hope.

NOTES.

Note A, page 6.

DECISIONS OF CRITICISM.—It is highly important to the student that the question concerning the true rank and esteem of the various styles and departments of painting, and of that of colouring in particular, should be rightly understood, that he may not waste his time and talents in vain and unprofitable endeavours; and, to comprehend this question correctly, it is necessary to consider these styles and departments in their true, natural, and philosophical relations individually, as well as in reference to art and society generally. In the first respect, they are either *material and mechanical—sensible and sentimental*—or they are *moral and intellectual*; yet it is impossible to separate these entirely from either branch, or to assign either to either otherwise than by predominance, in which way *hand, execution, drawing* and whatever concerns the management of the materials of a picture belong to the material and manual;—*colour, light, shade, and effect*, belong to sense,—and to the intellectual belong *invention, composition and expression*. Admitting, then, that intellect is above sense, and sense above matter and mechanism, we must accord the highest rank to invention and expression in themselves, and assign the like middle station to colouring; but if the art as a whole have, as it truly has, essentially more of the sensible than it has of either the material or intellectual, then is the sensible principal in the art, and colouring and its allies are principal in painting, although out of the art itself they have not the highest reference.

Styles in art have ever varied with their age and nation according as the people among whom they have been practised have been more or less mechanical, sensible, or intellectual,—without this, art would have terminated in rules, models, and sinensian sameness;—without this there had been neither Dutch, Flemish, nor Venetian schools; and without this there can be no English school. Had the critics and professors of Holland and Flanders inflated the Dutch and Flemish painters with the ambition of rivalling the schools of Michael Angelo, Raphael, and the Caracci, their works would have been rejected by their country, and their attempt to lead their age, instead of leaving art to the impulse of genius and the calls of society, would have had the effects it has had in England of neutralizing patronage or demand, and of depriving the world of a new developement of art, and of the rich produce of a school low in grade, but admirable in effect. Hence, left to its natural course in a free, great, and enlightened country like this, the art cannot fail of those new and transcendant attainments, which have ever sprung from power and riches, accompanied by freedom and intelligence, as witnessed in antient Greece; but the spirit of art in this country has been constantly depressed by false criticism, foreign revilings against its genius, and the domestic outcry of want of patronage for works of magnitude; as if magnitude were synonymous with merit. The Greeks knew how to concentrate more of the perfections of art on a gem than all the gigantic figures of Egypt, together, could afford;—why, then, shall not the British artist obtain riches and exalted reputation by the free exercise of his talents on works of such a magnitude as may be suited to the customs, climate and calls of his country?

Note B, page 7.

That Sir Joshua Reynolds felt unaffected upon a first inspection of the works of Raffael at Rome, was owing, doubtlessly, to the general unattractiveness of their colouring. Many, he observed, experienced the same indifference toward them. He remarked also, " That those persons only, who, from natural imbecility, appeared to be incapable of relishing those divine performances, made pretensions to instantaneous raptures on first beholding them."—Reynolds's Works by Farrington.

Gainsborough, with a candour parallel to that of Reynolds, acknowledged to Edwards, upon viewing the Cartoons at Hampton Court, that their beauty was of a class he could neither appreciate nor enjoy.

The present highly talented President of the Royal Academy has remarked, with just discrimination, that " They who have excelled in subjects of a grand and elevated character, have rarely been able to combine with their other accomplishments the merits of colouring, chiaro'scuro, and execution; but let us not, therefore, contract our ideas of excellence, in compliment to their deficiencies, nor endeavour to persuade ourselves that we see in the imperfection of their art a principle of their science."—Elements, Canto v. N. P. 284.

" How colouring and effect may and ought to be managed, to enliven form and invigorate sentiment and expression (remarks Opie), I can readily comprehend, and, I hope, demonstrate; but wherein these different classes of excellence are incompatible with each other, I could never conceive; nor will the barren coldness of David, the brickdust of the learned Poussin, nor even the dryness of Raffael himself, ever lead me to believe, that the flesh of heroes is less like flesh than that of other men; nor that the surest way to strike the imagination and interest the feelings, is to fatigue, perplex, and disgust the organs through which the impression is made on the mind."—Lect. 1. p. 18.

Upon this subject, and upon colouring in general, there has been much ably and eloquently said by Opie,—himself an eminent colourist,—in his fourth and last excellent lecture, which well merits the attention of every colourist, for just feeling and discrimination in this branch of painting.

It is evident, notwithstanding, that he was not well acquainted with the relations upon which harmony depends, since he confounds *tone and warmth* with harmony, according to a very common error, when he says, " Harmony is secured by keeping up the same *tone* through the whole, and *not at all by any sort of arrangement.*"—P. 143.

And again, " *Harmony* easily slides into *jaundice.*"—Ibid.

Now, harmony of colouring is infinite in its varieties, all depending upon *arrangement*, and tone is but the ruling colour pervading any arrangement or composition—the archeus of the piece; which in like case the harmonist, in the sister art of music, calls the key;—and warmth is the suffusion of a particular tone, the natural key-note of the colouring. Colouring has as many artificial keys as there are hues;—as many tones, all applicable to legitimate arrangement, or harmonies in colouring; but not all equally eligible, for this must depend upon taste, nature, sentiment, and judgment. Opie is not, however, singular in confounding tone with harmony; the error is general, and Sir Joshua has fallen into it;—but monotony of any kind must not be confounded with harmony.

Note C, pages 20. 226.

GREEN, ORANGE, VIOLET, AND INDIGO PRIMARIES — FROM CROSSINGS OF BLUE, RED, AND YELLOW RAYS.—Newton considered the simple originality of the first four of these colours, together with the three latter, to be demonstrated, because, after having produced them from a beam of light analytically, by prismatic refraction, he could not farther resolv either of them into other colours by passing them through a second prism; for, though variously dispersed, each colour retained its original hue.* Hence Newton concluded that there were seven primary colours.

Nevertheless, three of these, blue, red, and yellow, being separated from the others and properly compounded, reproduce colourless light, which again, by prismatic refraction, affords Newton's seven primaries, including green, orange, violet, and indigo; and so on repeatedly. †

Another difficulty into which Newton's hypothesis led, was the necessity of admitting two kinds of colours, which he denominated *homogeneal* and *heterogeneal*; thus his prismatic green was homogeneal, but a green composed of blue and yellow he called heterogeneal; yet, had he mixed his blue and yellow rays, *his prism* would have refracted without separating them, and thus the heterogeneal colour would have become homogeneal; we shall not, however, continue an argument which will make artists smile and grave men frown; yet, neither false shame, nor respect for authority, however exalted, ought, in any case, to suppress our regard for and vindication of truth. In the present case, too, that great man, Newton himself, admonishes us *not to admit more causes of natural appearances than are both true and sufficient to explain them*, agreeably to that more antient maxim, that *nature does nothing in vain*, and, therefore, would not have instituted seven primaries in a design which is perfectly accomplished by three.

Note D, pages 21. 29. 154. 249.

Having deduced the relations of colours regularly from white or light, through the primaries, secondaries, and tertiaries, to black or shade, we might have done the same inversely from black to white. On this plan the tertiaries, olive, russet, and citrine, take the place of the primaries, blue, red, and yellow, while the secondaries still retain their intermediate station and relation to both; thus russet and olive compose or unite in dark *purple*, citrine and olive in dark *green*, and russet and citrine in dark *orange*, as demonstrated pp. 248 and 249. The tertiaries have, therefore, the same order of relation to black that the primaries have to white; and we have black primaries,

---

* Optics, Prop. II. Theor. II. Exp. 5.   † Exp. XXVII. p. 247.

secondaries, and tertiaries, inversely, as we have white primaries, secondaries, and tertiaries, directly; or, what is the same thing, we have light and dark colours of all classes.

Theoretically, the tertiaries may be produced either by mixture of the primaries alone, the secondaries alone, or by the primaries and black; but in the latter mode the black must be *perfectly neutral* and the colours *true*, to do this practically, and none of our pigments are perfect enough for this; the latter mode is, therefore, a bad practice, and applicable only to the production of shadow colours, distinguished by the term semi-neutrals.—P. 28. It is the imperfection or anomalousness of pigments which renders these distinctions necessary, for had we pigments in the chromatic and relative perfection which belongs to prismatic colours, with also a perfectly transparent and neutral shade-colour with which to combine the whole, the inverse order of our classification would afford us the series from black to white; the contrary order adopted is, however, practically preferable, because we have white pigments sufficiently opaque and pure to compound all tints without changing the denominations of colours; but, as before remarked, we have no black so transparent and neutral as to afford us equally perfect shades: both these orders are, however, represented by the definite scale and the scale of equivalents taken conversely, and the absolute completeness of the natural system of colours is demonstrated analytically and synthetically, or rather antithetically.

Note E, page 22.

Mr. Brockedon, in his late interesting discussion of chromatic and optical phenomena at the Royal Institution, introduced a variety of devices, some of which, by very ingenious arrangement and mechanism, illustrated the combinations, contrasts, and powers of colours upon each other. To do justice to these by description, unassisted by engravings and colours, will be impossible; the following may, however, help the reader in forming a conception of one of them, which consisted of a broad ring, formed upon a white ground, and coloured at equal distances, blue, red, and yellow; these colours, being gradated alternately into each other, form the three secondaries, each opposite to its contrasting primary in regular gradation and series all round the ring or circle. A similar narrower circle of the mean diameter of the above, cut from its ground, internally and externally, formed a ring, which being coloured exactly as the above, but much more faintly, and placed concentrically upon it, *colour to colour*, alike all round, the power and mass of colour in the large ring negated that of the small incumbent ring, and rendered it nearly colourless to the eye; but the latter being turned half round, so that the colours of the two rings became opposed, *colour to contrast*, all round the circles, the faint hues of the small ring revived to the eye, with a vigour of colour far exceeding their natural power when viewed alone.

A colourless circular board, of sufficient diameter to cover the larger circle, being perforated with two or more small openings opposite to each other, and turned round concentrically upon this larger coloured ring, exhibited successively all the individual contrasts of the whole circular series, with other pleasing and instructive effects applicable to many uses in the arts.

This portion of Mr. Brockedon's plan is evidently available for our scale of equivalents, so as to isolate single contrasts, or by three or more openings, at equal distances, to indicate a like number of harmonizing colours, and the proportions on the scale in which they would compose harmoniously in painting.

Many diagrams have been contrived for exhibiting colours under various references; the schemes of Kircher, Lamozzo, Newton, and Harris, are well known. Adopting the doctrine of three primaries, &c. their relations may be illustrated by a great variety of trine figures; of which we have in our "Chromatics" preferred the triangle as the simplest and most analogous for form and composition.

Mr. Clover, who has studied this branch of his art with care and success, has constructed a diagram of this kind equally simple and ingenious, the principal device of which consists of a broad ring (circumscribing a triangle) constituted of three crescents, formed by lines drawn by the compasses from the outward to the inward edges of the ring, and dividing it into three equal portions, coloured severally blue, red, and yellow. It is evident that any diametrical line, drawn across this ring, will point to contrasts which arise from combinations of the primaries all round the ring.

Many other plans might be described, such as Mr. T. Hargreave's, which coincides with our own; Mr. Martin's, by the trisecting of three sets of concentric circles; and Mr. Hayter's, who has published a very curious and ingenious one, consisting of the involution of three spirals;— the principle of all which arises out of the admirable geometrical property of the apt combination of trine figures.

Note F, page 39.

Our theory of the constitution of light, upon which we interpret the phenomena of inherent and transient colours by chemical election, is fully adequate to the explanation of the colours of transmitted and reflected light. Thus, when light passes through transparent coloured glass, it is not the colour of the glass that tinges the light so transmitted, in the manner a colouring substance tinges a liquid, but the colour in the glass neutralizes itself, or retains from the light by election such a proportion of its principles as reduces such colour to an achromatic state, and suffers the remainder of the light to pass through the glass, so constituted as to afford the precise colour of the glass itself. In this manner such a proportion of the principles of light as constitutes redness passes through red glass, and the rest of the light is retained. It is the same with other colours of transparent substances; and in like manner the colours reflected from opaque coloured bodies are not the colours of the bodies themselves, but of the light by which they are illuminated—nor do we in any case see the immediate colours of objects, but those only with which they affect light. As yellow fixes or absorbs one proportion of the elementary principles of light—red another—blue another—and black absorbs the whole of light, so coloured bodies are found to become heated by the sun's rays in proportion as their colours retain or fix light, or refuse its transmission or reflection.

That colours and light itself are *oxides of hydrogen* is a doctrine which, though we have founded principally upon modern discoveries, is so remarkably coincident with one of a poetical and figurative character, drawn from traditions, and handed down to us by the father of poetry, that we may be pardoned perhaps for introducing it here. According to Hesiod, IRIS was the daughter of THAUMAS (or Osiris) and ELECTRA (or Isis), and trine sister of *Aello* and *Ocypete*.

What are we to understand by this? If *Iris* and her bow are figurative of COLOURS, *Electra* of the active principle of LIGHT, and *Thaumas* of the reactive principle of SHADE or darkness, as the learned will allow; and if *Ocypete* is also figurative of their offspring OXYGEN, and *Aëllo* [a stormy or gloomy air] also figurative of HYDROGEN, the reader will not find much difficulty in reconciling the poet's genealogy of Iris, or colours, with modern physics, and the chromatic theory we have founded upon them.

Could the true original significations of the names of the heathen deities be determined, the whole THEOGONY of Hesiod would probably unravel itself into the personifications of the powers of nature, according to the physics and metaphysics of the poet's time, and become as perspicuous to us as it must in such case have been to his contemporaries; or as his WORKS AND DAYS still are. This might account for the constant application of the heathen mythology to the allegories of poetry and painting, of which the personification of the powers of nature affords the grandest machinery, and sanctions the practice among Christians.

If this speculation is well founded, the painter may petition Iris with philosophic truth in the language of poetry, as imitated in the following Orphic, in which the colours are illustrated by natural objects, and classed in the order of their derivation, as primary, secondary, and tertiary.

TO IRIS.

Come, daughter of celestial right!
Come, goddess born of SHADE and LIGHT!*
Lovely and gay,—fair sister trine
Of Oxygen and Hydrogene,
Propitious come! and bring with thee
In glowing tints thy progeny.

Bring me the beam of *yellow* hue,
Bring me the sky's celestial *blue*,
   And Evening's roseate *red!*
Bring me fair Albion's vernal *green*,
The *orange* fruit, the *purple* vine,
   The fields with *citrine* spread.
O bring the *russet* heath, the *olive* grove,
And aid my genius in the art I love!

Note G, page 39.

UNEQUAL AFFINITIES OF THE ORGAN MAY EXPLAIN THE VARIOUS DEFECTS OF VISION WITH REGARD TO COLOURS.—An imperfect eye for colours is hardly perhaps less common than a bad ear for musical sounds. We were some years ago introduced to an intelli-

---

* Thaumas and Electra.

gent gentlemanly man, about fifty years of age, high in office at the East India House, who *never had been able to distinguish any colour* sufficiently to name it, nor could designate such otherwise than as comparatively *light* or *dark*, and who used to refer to his daughter, whose eyes were excellent, for their nominal distinctions. He used spectacles, but his eyes were not otherwise defective.

In the Philosophical Transactions for 1738 is an account of persons to whom all objects appeared *red* after having eaten henbane-roots.

In the same journal for 1777, p. 250, one Harris is mentioned who could not tell *black* from *white*. He had two brothers equally defective, one of whom mistook *orange* for *green*.

Again, (Phil. Trans. 1778, p. 613.) to another person full *reds* and full *greens* appeared alike, but *yellows* and *dark blues* were very nicely distinguished.

It is remarkable that in those cases of defective vision in which the eye is insensible to either of the primary colours, the party so defective confounds with, or regards such colour as, its opposite or contrasting colour.

This phenomenon is easily explained upon our principle of vision and colours, and the fact we have demonstrated in the preceding Experiments, III. VII. XII. XIII. that each colour contains virtually all the others; whence, if the eye be insensible to red, a red object even will appear to be *green*, and so of other cases. The rationale of a good or bad eye for colours rests upon the same ground, which depends upon the health or infirmity of the organ; and there can be no doubt that cultivation may very greatly improve the sensibility of the eye with regard to colours, as exercise strengthens the powers of the mind, and increases the health, vigour, and dexterity of the body: this is evident in the case of myopes, or the short-sighted, in whom the eye for colours is commonly deficient.

Inasmuch as a man is deficient in any sense, he is still unborn; whence those who are defective of vision are unconscious of, and never suspect, their own incapacity; yet the want of a good eye has not prevented some very eminent men from attaining high reputation in painting in spite of defective colouring; nor has it prevented others from investigating and writing on colours: nevertheless it is probable that much discordance in the theory of colours may have arisen upon this foundation. A late Professor of painting expresses himself upon this subject like one born blind or who had never seen colours, and many natural philosophers appear to have been remarkably deficient in this sense. Newton confesses that his eye for distinguishing colours was not very critical,[*] and that he availed himself of assistance in this respect. Mr. Dalton could not distinguish *blue* from *pink* by daylight.[†] Professor Sanderson, who was born, or very early became, blind, delivered lectures upon light and colours; and Dr. Priestley mentions an artist living in Edinburgh whose companions " have, by putting his colours out of the order in which he keeps them, sometimes made him give a gentleman a *green* beard, and paint a beautiful young lady with a pair of blue cheeks!"

Note H, page 40.

The effect of any colour intently viewed, in producing its opposite colour as an ocular spectrum; the effects of two colours of the prismatic spectrum, when cast separately into the two eyes

---

[*] Optics, Prop. III. Prob. I. p. 110.   [†] Manchester Mem. v. 28.

at the same time, in producing a compound sensation in the observer; the effects of colours contrasted contiguously in balancing or subduing each other by a similar combination; the like effects of transparent colours in glazing or mixture; the harmonizing influences of colours, and the whole doctrine of equivalence, are all attributable to the same principles.

Note 1, page 60.

As the eye of the artist is apt to be influenced in painting by surrounding colours, and as the same circumstance powerfully affects his finished works, the colour of the walls of the study, and gallery of the artist, and their accordance, are no unimportant aid or hinderance to good colouring; it has accordingly excited his attention with various results. The late academician Tresham, and his colleagues in the office of arranging the Townley collection of statues at the British Museum, found much difficulty in colouring the walls of the galleries in accordance with the best appearance of sculptures become dingy by age, owing to the well-known property of a plane surface, or mass of any particular colour, to obtrude or come forward upon the eye to the detriment of the relief of the statues, and the power of some hues to augment by contrast the foulness of their colour; both of which difficulties they ultimately overcame by sprinkling, or marbling, the walls with a second or third colour; upon which the walls retired from the eye, and the statues relieved, without any disadvantage from contrast, notwithstanding their having rather injudiciously adopted a warm advancing colour, better suited to a picture-gallery.

The principle which succeeded with the academicians in placing these statues has been carried into the painting-room and picture-gallery, perhaps irrelevantly and with ill effect, for pictures are in this respect opposed to statues, in which colour is of remote consideration, and relief principal. We look at a picture in its frame as if the representation had distance, and were seen through a window or door; the advancement or coming forward of the wall on which it hangs, so long as it does not attract the eye, is therefore a benefit rather than a disadvantage: consequently a plain colour or ground is preferable for hanging pictures upon; hence also frames of the boldest relief and most advancing colour exhibit pictures to the best advantage, by forming as it were a proscenium to the design, for pictures in this respect are dramas.

As to *colour*, those which are cold and dark are the most retiring; the warm and light advance most; and each colour has its antagonist, and consequently may affect a picture well or ill, according to its tone or general hue: hence there can be no universally good colour for the walls of a picture-gallery or painting-room; we may therefore conclude that a mean, or middle colour, not too obtrusive on the eye, is *generally* preferable; such is a *crimson* hue, compounded of a retiring and advancing colour, and neither hot nor cold,—which contrasts with the general *green* of nature and pictures. These are the middle colours of the chromatic system,—the most generally agreeable antagonists, and in almost all cases inoffensive to the eye.

We conclude, therefore, that a plain, flat, unobtrusive crimson colour is best adapted to the walls of an exhibition-room, and far superior to any other in general effect; it might also correct the too frequently prison-like appearance of the painting-room, and, if the mass of colour in this case should prove too advancing upon the eye of the artist, its power may be subdued by breaking it with a faint pattern or design.

Such a crimson will in general afford the most effective contrast to the works of the landscape-painter and subjects exhibiting distance, but is less essential to the portrait and historic painters, whose objects are more immediate and advancing; to such, therefore, a more retiring colour—a modest green, may in some cases prove more eligible; but the practice sometimes resorted to by the artist, of producing a favorable contrast for his pictures by a colour in itself disgusting upon his walls, is to be deprecated, as exciting an ill sentiment on entering the room by no means advantageous to himself or his works. In all cases, therefore, he should select a pleasing tint of colour; and, we may remark, that those of crimson and green are universally so, and that they are prime media of nature and art in effecting chromatic harmony: since however a universal rule cannot be given, the artist will have to exercise his judgment, according to the case, in selecting such hue as is best suited to the general character of his colouring, according to the principle of chromatic equivalence.

Upon this principle a bright fawn colour has been found by far the most favourable for contrasting the grey hue of the print in the hanging of engravings, and the only ground upon which they are viewed to advantage.

A cool gray, or neutral, is in general best suited to the passages and approaches of the gallery as a preparation of the eye, but is too retiring for the exhibition of pictures in general, although it is better suited to the sculpture which commonly ushers the visitor to the gallery or painting-room.

It might become a useful accessary to the study of an artist if sliding rods crossed the room diagonally, upon which a number of variously coloured and figured curtains moved beyond his subject or sitter, with which he might suit colours, or form combinations, draperies, &c. as backgrounds, or tune his eye upon feeling and principle to the colouring of his design. The utility and importance of appropriate backgrounds in portraiture, and even as auxiliaries to the rigid academic model, have been rendered so apparent by the precepts and practice of Sir Joshua Reynolds,[*] and they are so efficient in imparting meaning, sentiment, and harmony, to the otherwise inane and monotonous appearance of single figures, that they need hardly be urged in favour of such accessaries to the painting-room. The principle has indeed been acted upon of late years by some of those academicians who have been elected to the honourable distinction of directing the living school of the Royal Academy as visitors, and the practice must have proved eminently conducive to the progress of the student, to whom it supplied the means of fully comprehending the action, and the art of using the figure, while he traced with correctness its form; thus subjecting at once his hand, his eye, and his mind, to the same discipline. 'The art of seeing nature, or in other words the art of using models,' says Sir Joshua,[†] ' is in reality the great object—the point to which all our studies are directed.'

Form, and the simple figure, are, however, principal in sculpture, and in the rigid school of the living figure; nor should any accompaniments be allowed to infringe needlessly upon the time allotted for study, nor to run to the extreme of the minor schools; nor should the practice of the living school be confounded with the *tableau vivante* of the Continent, which has been refined upon and carried to perfection by Mr. Parris in the school of the Historical Society, wherein

---

[*] Sir J. R.'s Works, Note XLII.   [†] In his 12th Discourse.

almost every mode of composition and variety of accompaniment are judiciously introduced in life, light, shade, and colour, under the most beautiful and ingenious arrangements. Yet these several modes of practice have the same efficient principle, under the influence of which the pencil will acquire a fidelity preservative from false and unmeaning combinations, and a habit disposing the hand and eye to taste and effect, constituting a proper foundation for the poetry, expression, and sentiment—which are the offspring of feeling and intelligence, and the highest attainments of the art.

INDEX.

A.

Academic figure, *page* 263
Accidental colours 39, 236, 237
Acetate of Copper 130
           Lead 56
Aders, Mr., his collection of antient pictures, 2, 74
Adventitious colours 39, 236, 237
Aerial perspective 30
Affections expressed by colours 11, 14
African Green 130
Alcohol, its uses, 198, 200, 207
Almagra 95
Amber varnish 209
Analogy of painting and poetry 4, 15, 31
         colours and seasons 12
            colours and sounds 14, 31, 32, 227, 235, 247, 252
Analysis of light 35, 227
Anger, colour of, 11, 12, 89
Anotta 120, 186, 188
Antient Cyanus, 107
Antients, colouring of the, 1
Antimony Yellow 79, 119, 184, 188
Antwerp Brown 162, 187
        Blue 112
Argent, Blanc d', 69, 185
Armenian Blue 2, 107, 113
Arrangement of colours 15, 21, 28, 32, 61, 72, 85
Ashes, ultramarine, 110, 170
Asphaltum 161, 162, 187
Atticum, Sil, 95
Azure 106, 111

B.

Background, importance of, *page* 263
Bartholomew, Mr., colouring of, 97
Barytic White 70
Beauty of colours 5, 10, 53, 59
Bice Blue 113
     Green 130
Bistre 162, 187
Bitumen 161
Black Colour 30, 172, 225
     Lake 179
     Compound 179
     Bone 179
     Ivory 179, 187
     Lamp 180, 187
     Frankfort 180, 187
     Vine-twig 180
     Peach-stone 180
     Almond, &c. 180
     Blue 180
     Spanish, &c. 181
     Purple 181
     Mineral 181, 187
     Ochre, &c. 181, 187
     Chalk 181, 187
     Lead 182, 187
     Lead drawings, to fix, 182
Bladder Green 131
Blanc de Roi 71
     d'Argent 69, 185
Blood, why red, 87
     Dragon's 97, 186—188
Blue Colour 102
     Pigments 106, 184—192

INDEX.

Blue, Armenian, *page* 2, 107, 113
    Antwerp 112, 184, 186
    Cobalt 110, 185
    Royal 111, 185, 186
    Prussian 111, 184, 186
    Berlin 111
    Dumont's 111
    Saxon 111
    Hungary 111
    Enamel 111
    Vienna 111
    Paris 111
    Haerlem 112
    Indian 112
    Verditer 130, 185
    Bice 113
    Black 180
    Intense 113, 184
    Verditer 113, 185
    Saunders's 113
    Mountain 113, 185
    Schweinfurt 113
    Terre 113
    Ochre 114
    Carmine 114
Body of colours, what, 54, 61
Bole 95
Bongeval White 71
British School 5
Brockedon, Mr., his illustration of colours, 258
Brown, Mary-Anne, a good poetical colourist, 19, 88, 93, 150
Brown Colour 154
    Ochre 80
    Vandyke 159, 187
    Antwerp 162
    Manganese 160
    Rubens's 160, 187
    Spanish 161
    Campania 160
    Bone 161, 184
    Ivory 161
    Egyptian 162
    Field's 146, 163

Brown Chestnut, *page* 163, 187
    Lake 163
    Prussian 146, 163
    Madder 146, 163
    Intense 146
    Pink 142, 163, 184, 188
    Ink 163
Brunswick Green 130
Burnt Sienna Earth 119, 187
    Carmine 137, 184, 188
    Umbre 160
    Verdigris 152
    Roman Ochre 80, 187
Byron's poetic colouring 18, 76, 105, 121, 127

C.

Carmine 100, 184, 188
    Burnt 137
    of Madder 101
    Durable 101
    Blue 114
Carnations, Oil of, 206
Carucru 165, 186, 188
Cassel Earth 160, 187
Cassia Fistula 142, 184, 188
Cassius's Purple 136
Cendres Bleus 113
Chalk 71
Chemical constitution of light and colours 35
Chinese Yellow 81, 188
    Lake 99
    Vermilion 93
Chromascope 221, 223
Chromate of Mercury 118, 185
Chromatic equivalents 22, 238, 249
Chrome Yellows 51, 77
    Orange 118, 185
    Green 129, 187
Cinnabar 93
Citrine Colour 139
    Composed 141
    Pigments 142
Cleaning, picture, 216

INDEX.

Clover, Mr., his scheme of colours, *page* 259
Cobalt Blue 51, 110, 111
    Green 129, 187
Cologne Earth 160, 187
Colour, female eye for, 19
Colours, fundamental scale of, 21
    Relations of 20
    Powers of 21
    Perspective of 30
    Physical cause of 34, 259
    Semineutral 29, 154
    Expression of 11, 15, 62, 73, 87, 103, 116, 123, 133, 140, 144, 149, 156, 167, 174
    Male and female 53
Coloured Inks 99, 163
Colouring, Beauties of, 10
    of the Antients 1
        Moderns 3, 256
    Venetian 3, 4
    British 5
    Vicious extreme in 9
    Practical maxims of 41, 46, 48, 108
    Correggio's 48
    Rubens's 29, 41, 48
    Reynolds's 25, 47
    Titian's 25, 49
    Wilson's and Gainsborough's 49
    Last attained in painting 8
    Poetical 4, 5, 10, 15, 58, 63, 73, 87, 104, 116, 124, 133, 145, 149, 156, 164, 167, 173
    Of Shakspeare 6, 15, 61, 64—67, 72, 74, 75, 88—92, 106, 116, 124—127, 134, 145, 150, 154, 158, 167—169, 175, 176, 178
    Akenside 11, 16, 64, 65, 105, 127, 133, 151
    Coleridge 12
    Collins 15, 48, 104, 106, 125—127, 140, 151, 157, 158
    Spenser 15, 18, 63, 66, 74, 76, 89, 91, 92, 106, 125, 134, 148, 169, 175
    Prior, 43, 88, 106

Colouring, poetical, of Shee, *page* 44, 116, 196, 221
    Of Milton 6, 64, 67, 85, 88, 89, 91, 104, 106, 116, 126—128, 132, 134, 140, 145, 151, 152, 156, 157, 159, 168, 169, 175—177
    Rogers 64, 67, 116, 145, 178
    Pope 64, 67, 88, 89, 116, 145
    Addison 5, 64, 66, 125, 127
    Dryden 64, 74, 89, 105, 106, 124, 134, 145, 168
    Mason 65, 67, 159
    Horace 4, 65
    Middleton 65
    Richardson 66
    Butler 73, 90, 117, 141
    Chaucer 73, 76, 89, 124, 139, 140, 174
    Chapman 74
    Mary-Anne Brown 19, 88, 93, 150
    Byron 18, 76, 102, 106, 121, 127, 158, 168
    Haller 89
    Sir Wm. Jones 90
    Marlowe 90, 178
    Crashaw 91, 116
    Charlotte Smith 104, 125
    Fletcher 106
    Thomson 106, 115, 134, 157, 159, 177
    Mrs. Pickersgill 106, 117
    Ossian 106
    Rowe 116, 141
    Cowper 122
    Burns 124
    Virgil 125
    Quarles 126
    Gray 127
    Drummond 127
    Fenton 134
    Joanna Baillie 144
    Drayton 145
    Goldsmith 76, 158, 159
    Miss Seward 158
    Walter Scott 158

Colouring, poetical, of A. Cunningham *page* 169
    Sydenham 175
    Duncombe 176
    Young 178
Colouring, Vices of, 9, 25, 29, 41, 46, 59, 74, 87, 133, 167, 174
Colourist, pleasures of the, 10, 22
Compensatory colours 22, 238, 249
Complementary colours 22, 238, 249
Compound Pigments 25, 47, 117, 128, 136, 141, 146, 152, 159, 165, 170, 179
    Orange 117
    Green 128
    Purple 136
    Citrine 141
    Russet 146
    Olive 152
    Brown 159
    Marrone 165
    Gray 170
    Black 179
Constable, Mr., his practice in setting the figure, 263
    Chiaroscuro 174
    Remarks on Landscapes 123, 174
Constant White 70
Contrast, Principle of, 26, 247
    Doctrine of, 22, 26, 238, 247
Copal Varnish 203, 208
    Vehicle 208
Copper, Carbonate of, 113
    Greens, 129, 185
    Prussiate of, 146
Correggio's Practice 48, 213
Cousins's Tints 170
Crayon colours 192, 194
Cream White 69
Cyanide of Iron 112
Cyanus of the Antients 107

D.

Damonico, 119, 187
Diagrams of colours 21, 23, 224, 249, 258

Diana, her colour, *page* 104, 116
Distemper Painting 193
    Colours 192
Doctrine of contrasts 22, 26, 238, 247
Dragon's Blood 97, 186, 188
Drawings, Black-lead, to fix, 182
Dryades, their colour, 123
Dryers 56, 130, 160, 206
Drying Oils 56, 201, 204
    Prepared 205, 206
Durability of colours 44
Dutch School 4, 255
    Pink 84, 188

E.

Earth, Sienna, 80, 119, 187
    Cologne, 160, 187
    Cassel, 160
Eastlake, Mr., his setting of the figure, 263
Egg-shell White 71
Elements of light and colours 35
Emerald Green 130
Enamel Colours 191
    Blue 111
English Pink 84, 188
    Red 97
Engraver, colouring of the, 174, 178, 180
Engravings, &c. hanging of, 263
Equivalent Colours 22, 238
    Scale of 22, 249
Essential Oils 207
Etty, Mr., his practice in setting the figure, 263
Experiments on Light and Colours 221
    Coloured Light 234
    Shadow 239
    Moon and Star-light 241
Expressed Oils 201, 204
Expression of colours 11, 15, 62, 73, 87, 103, 116, 123, 133, 140, 144, 149, 156, 167, 174
Eye for colour 11, 15, 19, 39, 226, 236, 253, 261

Eye, influence of colour on, *page* 59, 73, 85, 102, 115, 132, 136, 238, 262
    Not cognisant of colours 236

F.

Falsalo 161
Female Eye for colour 19
Field's Lakes 97
    White Lac-varnish 204, 209
    Metrochrome 221, 244
    Chromascope 221, 223
    Colours 70, 81, 82, 97, 98, 101, 106, 120, 129, 136, 137, 142, 146, 153, 163, 165, 181
Figure, Academic, 263
Fire, its effects on colours, 44, 45
Fish Oils 206
Flake White 68, 69, 185
Flemish School 4, 255
Flesh, colouring of, 48
    Tints 48, 118
Flora, her colour, 173
Florentine Lake 93
Frankfort Black 180
French Green 130
Fresco 192, 193
Fundamental Scale of Colours 21
Furies, their colour, 173

G.

Gainsborough's practice 49
Gallery, how best coloured, 262
Gall-stone 83, 188
Gamboge 82, 83, 188
Garance, Bleu de, 110
    Laque de 98
Giallolini 78
Girtin's practice 132
Giulio Romano's Phaeton 9
Glazing and Scumbling 46, 48, 55, 190, 205

Gold, Purple of, *page* 136, 187
Gold-size, Japanners', 98, 206
Golden Sulphur of Antimony 119, 184, 188
Graphite 182, 187
Gray, Compound, 170
Green, Compound, 128
    Varley's 128
    Italian 128
    Hooker's 128
    Verona 129
    Brunswick 130
    Chrome 129, 185
    Cobalt 129
    Copper 129
    Mineral 130, 185
    Bice 130
    Scheele's 131
    Schweinfurt's 131
    Emerald 130, 131
    Mountain 130, 185
    French 130
    Olympian 130
    Persian 130
    Saxon 130
    Patent 130
    Marine 130
    Prussian 131
    Sap 131, 184
    Venetian 131
    Verditer 130, 185
    Bladder 131
Grief, colour of, 11, 13, 174
Gum 197
Gumtion 202, 208

H.

Haerlem Blue 112
Hamburgh Lake 99
Hanging of Pictures 262
    Engravings 263
Harding's Tints 170
Hargreave, Mr., his scheme of colours, 259

Harmony of Colouring, *page* 10, 14, 16, 20, 22, 27, 31, 51, 59, 61, 86, 103, 116, 122, 133, 144, 149, 150, 155, 167; 174, 227, 230, 238, 249, 256
Harpies, their colour, 173
Heat, effects of on colours, 44
Hewlett, Mr., colouring of, 97
History of Colouring 1
Hogarth's practice 47
Holy Green 123
Homer's poetic colouring 18, 59
Hooker's Green 128
        Mrs., Vehicle, 199
Hope, colours of, 13, 87, 88
Hues and Shades 21
        Shades and Tints distinguished 28
Hungary Blue 111
Hydro-sulphuret of Antimony 119, 184

I.

Imperial Purple 133
Indian Yellow 83
        Blue 112
        Red 96
        Lake 100
        Ink 182
Indigo 112, 184, 188
Ink, Indian, 182
        Red 98
        Brown 163
        China 182
Instruments, New Optical, 221, 243, 244
Intense Blue 113
        Brown 146
Invisible Green 131, 152
Iodine, Scarlet, 94, 186, 188
Iron, Cyanide of, 111
        Phosphate of, 114, 171
        affects Pigments 189
Italian Pink 84, 188
        Green 128
Ivory Black 179
        Brown 161

Japanners' Gold-size, *page* 98, 206
Jaune Minerale 77, 185
        De Mars 119, 187
Jealousy, colour of, 12, 74, 126
Jews' Pitch 161
Joy, colour of, 13, 87, 88
Juno, her colours, 104
Jupiter, his colour, 133

K.

Kermes Lake 99
King's Yellow 81, 188
Kircher's Scheme of Colours 259

L.

LacVarnish 204, 209
        Vehicle 199
Lake, Yellow, 83, 84, 184, 188
        Red 97
        Rubric 97
        Madder 97
        Scarlet 99, 184, 188
        Florentine 99, 184
        Hamburgh 99, 184, 188
        Roman 99
        Venetian 99
        Kermes 99, 184
        Chinese 99
        Liquid Madder 98
        Lac 100, 184, 188
        Field's 97
        Cochineal 97, 99, 188
        Quercitron 84, 184
        Green 130
        Purple 137, 184
        Brown 163
        Citrine 142
        Olive 153
        Black 179
        Marrone 165
Lamp-black 180

INDEX.

Laque Minerale, *page* 118, 185
Lavender Oil 207
Lawrence, Sir Thomas, his practice, 19, 48, 96
Lead, White, 68, 185
    Black 182
    Chromate of 77
    Red 95
    Orange 119
    Its effects on Colours 188
    Sulphate of 69, 185
Lemon Yellow 81, 187
Lenses 223
Lensic Prism 222
Leslie, Mr., his practice in setting the figure, 263
Light, Analysis of, 35, 227
    how constituted 35
    a compound substance 36
    a principle of colours 37
    Action of on colours 184
    Perspective of 30
Light and Colours chemically constituted 35
    Experiments on 222
Light and Shade 35, 224
Linseed Oil 201, 204
    Drying 205
Liquid Rubiate 98
Local colour, what, 32
London White 68
Love, Colour of, 11, 13, 19, 85, 87, 88

M.

Macpherson's Tints 170
Madder, Yellow, 82
    Carmine 97, 101, 187
    Lake 97, 187
    Purple 137, 187
    Liquid 98, 187
    Brown 146, 187
    Russet 146, 187

Madder, Intense Brown, *page* 146, 187
    Marrone 165, 187
Magilp 202, 208
Malachite 131
Manganese Brown 160, 187
Marine Green 130
Marrone Colour 164
Mars, his colour, 87, 89
Martin, Mr., his scheme of colours, 259
Massicot 79, 188
Mastic Varnish 208
Maxim of Rubens 29, 41, 48
    Reynolds 25, 47
Measure of Colours 245, 246
Melinum 71
Melos White 71
Mengs 81
Mercury, Submuriate of, 70, 186
    Chromate of, 118, 185
Metamorphosis of Colours, 234
Metrochrome 221, 244
Michael Angelo 8
Milton, a poetic colourist, 18
Mineral Turbeth 78
    Black 181
    Green 130
    Purple 137
    Pitch 161
Minerale, Jaune, 77
Minerva, her colour, 104, 105
Minium 93, 95, 119
Mixed Colours 21, 25, 47, 117, 128, 136, 141, 146, 152, 159, 165, 170, 179
Modan White 71
Monochrome 252
Montpellier Yellow 77
Moral Colouring 12, 63
Morat White 71
Mountain Green 131
Mucilage 198
Mummy 162, 187
Music and Colouring, analogy of, 31, 227, 235, 247, 252

N.

Naiades, their colour, *page* 123, 125
Naples Yellow 78
Nature, the source of colouring, 15
Neptune, his colour, 123
Neutral, The, how constituted, 21
    Tint 170
    White 61
    Black 172
    Grey 166
Newton's seven primary colours, how redundant, 225, 257
Nottingham White 68
Nut Oils 206

O.

Ochre, Yellow 79
    Stone 79
    Roman 80
    Brown 80
    Spruce 80
    de Rue 80
    Red 95
    Scarlet 95, 96
    Indian 95
    Blue 114, 187
    Orange 119
    Purple 137
Ocular Spectra explained 22, 39, 236, 237
Oil, Expressed 201, 204
    Olive 206
    Poppy 206
    Volatile 207
    Linseed 201, 204
    Nut 206
    of Lavender or Spike 207
    Pinks 206
    Turpentine 207
    Drying 205, 206
    Fish 206
Oils, Essential or Volatile, 207
Olive Colour 148

Olive Compound, *page* 152
    Green 152
    Lake 153
Opacity, what, 37, 54, 61
Orange Colour 115
    Vermilion 118, 187
    Chrome 118, 185
    Orpiment 119, 184, 188
    Compound 117
    Ochre 119, 187
    Lead 119, 185, 188
    Lake 120
    Russet 146, 187
Orpiment, Yellow, 80, 188
    Orange, 119
    Red, 119
Owen's practice 48
Ox-gall 200
Oxford Ochre 79
Oyster-shell White 71

P.

Painting, Schools of, 4, 255, 263
    Maxims of, 25, 29, 41, 47, 48
    Principle of Practice in 47
    in Distemper 193
    Crayon 192, 194
    Oil 200
    Fresco 192, 193
    Water 197
Painting-Room, colour of, 262
Palette, Setting of the, 32
Paris Blue 111
    White 71
Parris's School of Painting 263
Passions expressed by colours 11, 14, 74, 87, 104
Patent Green 130
    Yellow 77
Pearl White 70, 187
Permanence 44, 187
Permanent White 70
    Pigments 187

INDEX.

Persian Red, *page* 96
    Green 130
Perspective, Aerial, 30
    of Colours 30
Phosphate of Iron 114, 171, 187
Picture-Cleaning 216
Pictures, Hanging of, 262
Pigments, General Qualities of, 53
    White 68
    Yellow 77
    Red 93
    Blue 106
    Orange 117
    Green 128
    Purple 136
    Citrine 141
    Russet 146
    Olive 152
    Brown 159
    Marrone 165
    Gray 170
    Black 179
    Tables of 183
    action of light on 184
    affected by impure air 185
    imperfect 186
    most permanent 187
    acted on by lead 188
    acted on by iron 189
    transparent 190
    unaffected by heat 191
    unaffected by lime 192
Pink, English, Dutch, and Italian, 84, 188
    Madder Lake 97
    Brown 142, 163
Pinks, Oil of, 206
Platina Yellows 81
Plumbago 182
Poetic Colouring 4, 10, 15, 58, 63, 73, 87, 104, 116, 124, 133, 145, 149, 156, 164, 167, 173
Poetry and Painting, analogy of, 4, 15, 31
Poets, good or bad colourists, 15
Poppy Oil 206
Poussin's Deluge, Colouring of, 9
Powers of Colours 22, 61, 72, 85, 183, 248

Precipitate, Cassius's, *page* 136
Primary Colours 20, 72, 85, 102, 225, 257
Principle of Contrasts 26, 247
Prismatic spectra 39, 224, 228, 240
Prismic Lens 222
Prisms 222
Process, Indian 199
    Venetian 198
    Mr. Robertson's 198
    Mrs. Hooker's 199
    Mr. Jones's 200
Prussian Green 131
    Red 97
    Blue 111
Prussiate of Copper 146, 186
Purple Colour 132
    Tyrian 133
    Compound 136
    of Gold 136, 187
    Madder 137
    Lake 137, 188
    Ochre 137
    Mineral 137
    Black 165, 181

Q.

Qualities of Pigments 53, 183
Quercitron Lake 84, 142
Quicksilver, Sulphuret of, 93, 118, 187

R.

Raffael's practice 3
Rainbow 233
Raw Umbre 143, 187
    Sienna Earth 80
Recipe for Drying Oils 56
    Drying Lakes 98
    Purifying Gamboge 83
    Composing Colours 21, 25, 47, 117, 136, 141, 146, 152, 159, 165, 170, 179
    Hooker's Vehicle 199
    Robertson's Ditto 198
    Jones's Ditto 200

274   INDEX.

Recipe for Japanners' Gold Size, *page* 206
Red Colour 85
    Chrome 95, 185
    Lead 95, 185, 188
    Ochre 95, 187
    Indian 96, 187
    Saturnine 95
    Light 95, 96, 187
    Venetian 95, 96, 187
    Persian 96
    Prussian 97
    English 97
    Spanish 97
    Orpiment 101, 119
Relations of Colours 20
Reynolds's, Sir J., practice 25, 46, 47, 49, 81, 82, 167
Roman Lake 99
    Ochre 80, 187
    White 68, 69
Romano's, Giulio, Phaeton 9
Rose Pink 101, 184
    Madder Lake 97, 187
Roucou 120
Rouen White 71
Rouge 101, 165, 184
Royal Blue 111, 186
Rubens, Maxims of, 29, 41, 48
    Practice of 47
    Brown 160
Rubiate Liquid 98, 187
Rubric Lakes 97, 187
Rue, Ochre de, 80
Russet Colour 144, 187
    Compound 146
    Rubiate 146, 187
    Orange 146, 187
    Purple 146

S.

Sanguis Draconis 97
Sap Green 131, 184, 188
Satin White 71
Saturnine Red 95

Saunders's Blue, *page* 113
Saxon Green 130
    Blue 111
Scale of Colours 21, Pl. 1. fig. 3
    of Equivalents 23, Pl. 1. fig. 2
Scarlet Iodine 94, 186
    Ochre 96
    Lake 99
Scheele's Green 131
Schemes of Colours 258
    The Author's 21, 23, 224, 249, 259
    Mr. Brockedon's 258
    Mr. Clover's 259
    Mr. Hargreaves's 259
    Harris's— Hayter's — Kircher's — Lamozzo's—Martin's—Newton's 259
Schools of Painting 4, 255, 263
Schweinfurt Blue 113
    Green 131
Sculpture Gallery 263
Secondary Colours 21, 115
Semineutral Colours 29, 154
Sensorium, Colours belong to the, 237
Sepia 162, 187
Shade, Analysis of, 224
Shade and Light 35, 41, 174, 224, 239, 258
Shades, Hues, and Tints 28
Shadow, Colours of, 239
Shakspeare a great colourist 16
Sienna Earth 80
    Burnt 119
Sil Atticum 95
Silver White 69
Sky, Colours of, 103, 166, 233, 239
Smalt 111, 185
Sorrow, Colour of, 11, 13, 174
Sounds and Colours, Analogy of, 31, 227, 235, 247, 252
Spanish Brown 161
    White 71
Spectra, Ocular 39, 236, 237
    Prismatic 39, 224, 240
Spencer a poetical colourist 63, 125
Spike, Oil of, 207
Spirit of Wine 198, 200, 207

Spruce Ochre, *page* 80
Standard of Colours 244
Stil de Grain 84
Stone Ochre 79
Sub-Phosphate of Iron 114
Sulphate of Lead 69
Sulphuret of Mercury 93, 118
    Antimony 119, 186
    Arsenic 80, 188
Sympathy and Antipathy of Green 54, 121

T.

Tableau Vivante 263
Tables of Pigments 183
    affected by light, &c. 184
    by impure air, &c. 185
    by lead 188
    by iron 189
    unaffected by heat 191
    by lime 192
    transparent 190
    imperfect 186
    most permanent 187
Teniers' practice 47
Terre Bleu 113
  Verte 128
  de Cassel 160
Tertiary Colours 21, 139, 257
  Composed 21
Test of Ultramarine 109, 110
  Lakes 98
  Vermilions 94
Theory and Practice 32, 47, 104, 123, 140, 250, 263
Time, its Effects on Colours, 44
Tin, White 70
Tints 28
  Macpherson's 170
  Harding's 170
Titian's practice 3, 8, 25, 68, 213
Tone and Harmony differ 256
Transient Colours 36
Transparency and Opacity 37, 55, 61, 62, 259

Transparent Pigments, *page* 190
Trees, Colours of, 122, 157
Triunity of colours, indivisible, 20, 31, 72, 85, 102, 225, 227, 235, 247, 252, 257
Troy White 71
Turbeth Mineral 78, 185, 186
Tyrian Purple 133

U.

Ultramarine 106, 187
  Ashes 110, 170, 187
  Antiquity of 107
  Test for 109, 110
  Factitious 110
  Dutch 111
  False 111
Umbre Burnt 160
  Raw 143

V.

Vandyke's practice 9, 25, 48, 109, 159
  Brown 159
Varley's Green 128
Varnish, Mastic 208
  Copal 203, 208
  Amber 209
  White Lac 204, 209
Varnishes 208
Varnishing 203, 209
Vehicle, Copal, 208
  Mastic 202
  Lac 199, 209, 210
  Hooker's 199
  Jones's 200
  Robertson's 198
  Venetian 198
  Indian 199
Vehicles 196
  Water 197
Venetian Process 198
  Grounds 213
  School 4, 255
  Red 95, 96
  Green 131

Verdigris, *page* 130, 185, 186
    Burnt 152
Verditer Blue 113
    Green 113
Vermilion Red 93, 187
    Orange 118, 187
Verona Green 129
Veronese's, Paul, Grounds, 213
Verte, Terre, 128, 187
Vices in Colouring 9, 25, 29, 41, 46, 59, 74, 87, 133, 167, 174
Vienna Blue 111
Viride Æris 130
Vision, New Theory of, 39, 236
    Powers of Colour on, 59, 73, 85, 102, 115, 132, 136, 238, 262
Vitrified Pigments 45
Volatile Oils 207

W.

Water Vehicles 197
West's Practice 82, 96
White Colour 61
    Lead 68, 185
    Nottingham 68
    London 68
    Crems 69, 185
    Pearl 70, 186
    Tin 70, 187
    Zinc 70
    Silver 69
    Flake 69, 185
    Antimony 70, 186
    Satin 71, 187
    Melos 71, 187
    Modan 71, 187
    Morat, 71, 187

White Bismuth, *page* 70, 186
    Bougeval 71
    Egg-shell 71, 187
    Spanish 71, 187
    Oyster-shell 71, 187
    Troy 71
    Roman 69, 185
    Rouen 71
    Venetian 69, 185
Wilkie, Mr., his practice in setting the figure, 263
Wilson's Practice 31, 47, 49, 82
Wine, Spirit of, 198, 200, 207

Y.

Yellow Colour 72
    Pigments 76
    Chrome 51, 77, 185
    Patent 77, 185, 186
    Montpellier 77
    Naples 78, 185
    Ochre 79, 187
    Orpiment 80, 184
    King's 81, 184
    Chinese 82, 184
    Platina 83, 187
    Madder 82
    Indian 83, 184, 188
    Lake 83, 184, 188
    Lemon 81, 187

Z.

Zaffre 111
Zinc White 70, 187
    Sulphate of 56

THE END.

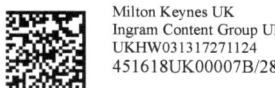
Milton Keynes UK
Ingram Content Group UK Ltd.
UKHW031317271124
451618UK00007B/284